HORACE SECRIST

READINGS AND PROBLEMS

IN

STATISTICAL METHODS

Elibron Classics
www.elibron.com

Elibron Classics series.

ISBN 1-4021-6848-9 (paperback)
ISBN 1-4212-8272-0 (hardcover)

This Elibron Classics Replica Edition is an unabridged facsimile of the edition published in 1920 by the Macmillan Company, New York.

READINGS AND PROBLEMS IN STATISTICAL METHODS

THE MACMILLAN COMPANY
NEW YORK · BOSTON · CHICAGO · DALLAS
ATLANTA · SAN FRANCISCO

MACMILLAN & CO., LIMITED
LONDON · BOMBAY · CALCUTTA
MELBOURNE

THE MACMILLAN CO. OF CANADA, LTD.
TORONTO

READINGS AND PROBLEMS

IN

STATISTICAL METHODS

BY

HORACE SECRIST, Ph.D.

PROFESSOR OF ECONOMICS AND STATISTICS
NORTHWESTERN UNIVERSITY

DIRECTOR, THE BUREAU OF BUSINESS RESEARCH
NORTHWESTERN UNIVERSITY SCHOOL
OF COMMERCE

AUTHOR OF "AN INTRODUCTION TO STATISTICAL METHODS"

New York

THE MACMILLAN COMPANY

1920

Set up and electrotyped. Published October, 1920.

Norwood Press
J. S. Cushing Co. — Berwick & Smith Co.
Norwood, Mass., U.S.A.

CONTENTS

INTRODUCTION

The selections included in this book were chosen to illustrate concretely the attitude of mind in which statistical analysis must be undertaken, and to develop logically the steps and processes through which statistical data must be carried in order to be used as bases for logical inferences. They constitute within themselves an independent treatment of statistical principles; but, undoubtedly, will have their greatest value when used in connection with a text on statistical methods. They are intended primarily to be used in this manner.

The *use* of statistics is consciously emphasized. "Embalmed" statistics have no part in the treatment, as they have no place in the writer's interest. The collection, use, and interpretation of statistical data are justified largely, if not solely, in the service which they have for planning, whether it is related to questions of social control, business policy, or statecraft.

It has seemed wise to accompany the selections with pertinent, thought-provoking questions, which students and others may use as a basis for criticism and constructive analysis. Accordingly, review questions are made component parts of the treatment. It is not intended that these shall be used solely as a means of making easy the assimilation of the contents of the selections, but rather, that they shall serve to connect the subject matter with the experience and training of the reader.

Review Problems have been added at the close of those chapters, the subject matter of which seems to lend itself to concrete application or to laboratory use. It is the teachers' obligation to make his laboratory exercises of interest to those whom he asks to take part in them, and to couple them with concrete business, industrial and social experiences. The *make-work* problems to which students are too often assigned, as part of their laboratory work, not only fail to arouse intellectual interest, but have the effect of divorcing the laboratory from the life which the student is living. They are too often looked upon as tasks or penalties, rather than as opportunities to take part in explaining, illustrating, and summarizing data which have to be manipulated before they can be used as bases for business and social judgments.

Laboratory problems should be chosen from business and social fields, and should include topics in which the student himself has an interest, and which he would be willing and eager to study statistically, in order more fully to understand. It is not difficult to select problems of this character and to secure data relating to them. In no other single problem, in the writer's experience as a teacher, has so much interest been developed on the part of his students in statistics, as in the study of expenditures for food at a local cafeteria. Theater tickets, types of business buildings, real estate valuations, show window decorations, classified advertisements, types of news items, stock and bond quotations, money rates, etc., all lend themselves to statistical treatment and arouse statistical interest. The writer has never been at a loss to find problems which create interest and which are worthy of study.

It is, therefore, with considerable hesitation that so-called *Review Problems* have been included in this book. The repeated requests on the part of instructors in Statistics for laboratory problems is the primary excuse which the writer has for including them here. It is hoped that they will be found of some interest to instructors in solving their laboratory difficulties, or of calling their attention to the problems immediately about them which may be used in their stead.

The frequent references to the *Text* in the *Reviews* and *Review Problems* are to the author's *Introduction to Statistical Methods*. While the *Introduction* and *Readings* are intended to be used together, either of them may be used separately for text or general purposes. It has seemed wise to employ the same chapter headings in the two volumes and this plan is followed. Chapters VI and VII, VIII, IX and X, XI, and XII in the *Introduction*, however, become Chapters VI, VII, VIII, IX, and X, respectively, in the *Readings*.

It is a pleasure for the writer to acknowledge his obligation to the authors and publishers of the selections included for the privilege of reprinting them, and to express his appreciation of the value which they have been to him in clarifying his own ideas on the meaning, function, and use of statistical methods in the understanding of business and social problems. It is the writer's hope that they will be equally interesting to those into whose hands this volume may come.

HORACE SECRIST.

NORTHWESTERN UNIVERSITY,
EVANSTON-CHICAGO, ILLINOIS.
June, 1920.

READINGS AND PROBLEMS IN STATISTICAL METHODS

CHAPTER I

THE MEANING AND APPLICATION OF STATISTICS AND STATISTICAL METHODS

Scientific Method — Its Scope and Meaning [1]

Within the past forty years so revolutionary a change has taken place in our appreciation of the essential facts in the growth of human society, that it has become necessary not only to rewrite history, but to profoundly modify our theory of life and gradually, but none the less certainly, to adapt our conduct to the novel theory. The insight which the investigations of Darwin, seconded by the suggestive but far less permanent work of Spencer, have given us into the development of both individual and social life, has compelled us to remodel our historical ideas and is slowly widening and consolidating our moral standards. This slowness ought not to dishearten us, for one of the strongest factors of social stability is the inertness, nay, rather active hostility, with which human societies receive all new ideas. It is the crucible in which the dross is separated from the

[1] Adapted with permission from Pearson, Karl, *The Grammar of Science*, Second edition, revised and enlarged, Chapter I, pp. 1–14. A. and C. Black, London.

genuine metal, and which saves the body-social from a succession of unprofitable and possibly injurious experimental variations. That the reformer should often be also the martyr is, perhaps, a not over-great price to pay for the caution with which society as a whole must move; it may require years to replace a great leader of men, but a stable and efficient society can only be the outcome of centuries of development.

If we have learned, it may be indirectly, from the writings of Darwin that the methods of production, the mode of holding property, the forms of marriage, the organizations of the family and of the commune are the essential factors which the historian has to trace in the growth of human society; if in our history books we are ceasing to head periods with the names of monarchs and to devote whole paragraphs to their mistresses, still we are far indeed from clearly grasping the exact interaction of the various factors of social evolution, or from understanding why one becomes predominant at this or that epoch. We can indeed note periods of great social activity and others of apparent quiescence, but it is probably only our ignorance of the exact course of social evolution which leads us to assign fundamental changes in social institutions either to individual man or to reformations and revolutions. We associate, it is true, the German Reformation with a replacement of collectivist by individualist standards, not only in religion but also in handicraft, art, and politics. The French Revolution in like manner is the epoch from which many are inclined to date the rebirth of those social ideas which have largely remolded the medieval relations of class and caste, relations little affected by the sixteenth-century Reformation. Coming somewhat nearer to our own time we can indeed measure with some degree of accuracy the social influence of the great changes in the methods of production, the transition from home to capitalistic industry, which transformed English life in the first half of this century, and has since made its way throughout the civilized world. But when we actually reach our own age, an age one of the most marked features of which is the startlingly rapid growth of the natural sciences and their far-reaching influence on the standards of both the comfort and the conduct of human life, we find it impossible to compress its social history into the bald phrases by which we attempt to connote the characteristics of more distant historical epochs. . . .

The contest of opinion in nearly every field of thought — the struggle of old and new standards in every sphere of activity, in religion, in commerce, in social life — touches the spiritual and physical needs of the individual far too nearly for him to be a dispassionate judge of the age in which he lives. That we play our parts in an era of rapid social change can scarcely be doubted by any one who regards attentively the marked contrasts presented by our modern society. It is an era alike of great self-assertion and of excessive altruism; we see the highest intellectual power accompanied by the strangest recrudescence of superstition; there is a strong socialist drift and yet not a few remarkable individualist teachers; the extremes of religious faith and of unequivocal freethought are found jostling each other. Nor do these opposing traits exist only in close social juxtaposition. The same individual mind, unconscious of its own want of logical consistency, will often exhibit our age in microcosm.

It is little wonder that we have hitherto made small way towards a common estimate of what our time is really contributing to the history of human progress. The one man finds in our age a restlessness, a distrust of authority, a questioning of the basis of all social institutions and long-established methods — characteristics which mark for him a decadence of social unity, a collapse of the time-honored principles which he conceives to be the sole possible guides of conduct. A second man with a different temperament pictures for us a golden age in the near future, when the new knowledge shall be diffused through the people, and when those modern notions of human relations, which he finds everywhere taking root, shall finally have supplanted worn-out customs.

One teacher propounds what is flatly contradicted by a second. "We want more piety," cries one; "We must have less," retorts another. "State interference in the hours of labor is absolutely needful," declares a third; "It will destroy all individual initiation and self-dependence," rejoins a fourth. "The salvation of the country depends upon the technical education of its work people," is the shout of one party; "Technical education is merely a trick by which the employer of labor thrusts upon the nation the expense of providing himself with better human machines," is the prompt answer of its opponents. "We need more private charity," say some; "All private charity is an anomaly, a waste of the nation's resources and a pauperizing of its members," reply others. "Endow scientific research and we shall know the truth, when and where it is possible to ascertain it"; but the counterblast is at hand: "To endow research is merely to encourage the research for endowment; the true man of science will not be held back by poverty, and if science is of use to us, it will pay for itself." Such are but a few samples of the conflict of opinion which we find raging around us. The prick of conscience and the spur of highly wrought sympathy have succeeded in arousing a wonderful restlessness in our generation — and this at a time when the advance of positive knowledge has called in question many old customs and old authorities. . . .

The state has become in our day the largest employer of labor, the greatest dispenser of charity, and, above all, the schoolmaster with the biggest school in the community. Directly or indirectly the individual citizen has to find some reply to the innumerable social and educational problems of the day. He requires some guide in the determination of his own action or in the choice of fitting representatives. He is thrust into an appalling maze of social and educational problems; and if his tribal conscience has any stuff in it, he feels that these problems ought not to be settled, so far as he has the power of settling them, by his own personal interests, by his individual prospects of profit or loss. He is called upon to form a judgment apart, if it possibly may be, from his own feelings and emotions — a judgment in what he conceives to be the interests of society at large. It may be a difficult thing for the large employer of labor to form a right judgment in matters of factory legislation, or for the private schoolmaster to see clearly in questions of state-aided education. None the less we should probably all agree that the tribal conscience ought for the sake of social welfare to be stronger than private interests, and that the *ideal* citizen, if he existed, would form a judgment free from personal bias.

Science and Citizenship

How is such a judgment — so necessary in our time with its hot conflict of individual opinions and its increased responsibility for the individual citizen — how is such a judgment to be formed? In the first place it is obvious that it can only be based on a clear knowledge of facts, an appreciation of their sequence and relative significance. The facts once classified, once understood, the judgment based upon them ought to be independent of the individual mind which examines them. Is there any other sphere, outside that of ideal citizenship, in which there is habitual use of this method of classifying facts and forming judgments upon them? For if there be, it cannot fail to be suggestive as to methods of eliminating individual bias; it ought to be one of the best training grounds for citizenship. The *classification of facts and the formation of absolute judgments upon the basis of this classification — judgments independent of the idiosyncrasies of the individual mind — essentially sum up the aim and method of modern science.*[1] The scientific man has above all things to strive at self-elimination in his judgments, to provide an argument which is as true for each individual mind as for his own. *The classification of facts, the recognition of their sequence and relative significance is the function of science,* and the habit of forming a judgment upon those facts unbiased by personal feeling is characteristic of what may be termed the scientific frame of mind. The scientific method of examining facts is not peculiar to one class of phenomena and to one class of workers; it is applicable to social as well as to physical problems, and we must carefully guard ourselves against supposing that the scientific frame of mind is a peculiarity of the professional scientist.

The First Claim of Modern Science

I have gone a rather roundabout way to reach my definition of science and scientific method. But it has been of

[1] The italics are not found in the original.

purpose, for in the spirit — and it is a healthy spirit — of our age we are accustomed to question all things and to demand a reason for their existence. The sole reason that can be given for any social institution or form of human activity — I mean not how they came to exist, which is a matter of history, but why we continue to encourage their existence — lies in this: their existence tends to promote the welfare of human society, to increase social happiness, or to strengthen social stability. In the spirit of our age we are bound to question the value of science; to ask in what way it increases the happiness of mankind or promotes social efficiency. We must justify the existence of modern science, or at least the large and growing demands which it makes upon the national exchequer. Apart from the increased physical comfort, apart from the intellectual enjoyment which modern science provides for the community . . . there is another and more fundamental justification for the time and energy spent in scientific work. From the standpoint of morality, or from the relation of the individual unit to other members of the same social group, we have to judge each human activity by its outcome in *conduct*. How, then, does science justify itself in its influence on the conduct of men as citizens? I assert that the encouragement of scientific investigation and the spread of scientific knowledge by largely inculcating scientific habits of mind will lead to more efficient citizenship and so to increased social stability. Minds trained to scientific methods are less likely to be led by mere appeal to the passions or by blind emotional excitement to sanction acts which in the end may lead to social disaster. In the first and foremost place, therefore, I lay stress upon the educational side of modern science, and state my position in some such words as these:

Modern Science, as training the mind to an exact and impartial analysis of facts, is an education specially fitted to promote sound citizenship.

Our first conclusion, then, as to the value of science for practical life turns upon the efficient training it provides in *method.* The man who has accustomed himself to marshal facts, to examine their complex mutual relations, and predict upon the result of this examination their inevitable sequences — sequences which we term natural laws and which are as valid for every normal mind as for that of the individual investigator — such a man, we may hope, will carry his scientific method into the field of social problems. He will scarcely be content with merely superficial statement, with vague appeal to the imagination, to the emotions, to individual prejudices. He will demand a high standard of reasoning, a clear insight into facts and their results, and his demand cannot fail to be beneficial to the community at large.

Essentials of Good Science

I want the reader to appreciate clearly that science justifies itself in its methods, quite apart from any serviceable knowledge it may convey. We are too apt to forget this purely educational side of science in the great value of its practical applications. We see too often the plea raised for science that it is *useful knowledge,* while philology and philosophy are supposed to have small utilitarian or commercial value. Science, indeed, often teaches us facts of primary importance for practical life; yet not on this account, but because it leads us to classifications and systems independent of the individual thinker, to sequences and laws admitting of no play-room for individual fancy, must we rate the training of science and its social value higher than those of philology and philosophy. Herein lies the first, but of course not the sole, ground for the popularization of science. That form of popular science which merely recites the results of investigations, which merely communicates *useful knowledge*, is from this standpoint bad science, or no science at all. Let me recommend the reader to apply this test to every work professing to give a popular account of any branch of science. If any such work gives a description of phenomena that appeals to his imagination rather than to his reason, then it is bad science. The first aim of any genuine work of science, however popular, ought to be the presentation of such a classification of facts that the reader's mind is irresistibly led to acknowledge a logical sequence — a law which appeals to the reason before it captivates the imagination. Let us be quite sure that whenever we come across a conclusion in a scientific work which does not flow from the classification of facts, or which is not directly stated by the author to be an assumption, then we are dealing with bad science. Good science will always be intelligible to the logically trained mind, if that mind can read and translate the language in which science is written. The scientific method is one and the same in all branches, and that method is the method of all logically trained minds. . . .

I would not have the reader suppose that the mere perusal of some standard scientific work will, in my opinion, produce a scientific habit of mind. I only suggest that it will give some insight into scientific method and some appreciation of its value. Those who can devote persistently some four or five hours a week to the conscientious study of any *one* limited branch of science will achieve in the space of a year or two much more than this. The busy layman is not bound to seek about for some branch which will give him useful facts for his profession or occupation in life. It does not indeed matter for the purpose we have now in view whether he seek to make himself proficient in geology, or biology, or geometry, or mechanics, or even history or folklore, if these be studied scientifically. What is necessary is the *thorough* knowledge of some small group of facts, the recognition of their relationship to each other, and of the formulæ or laws which express scientifically their sequences. It is in this manner that the mind becomes imbued with the scientific method and freed from individual bias in the formation of its judgments. . . .

The Scope of Science

The reader may perhaps feel that I am laying stress upon *method* at the expense of material content. Now this is the peculiarity of scientific method, that when once it has become a habit of mind, that mind converts *all* facts whatsoever into science. The field of science is unlimited; its material is endless, every group of natural phenomena, every phase of social life, every stage of past or present development is material for science. *The unity of all science consists alone in its method, not in its material.* The man who classifies facts of any kind whatever, who sees their mutual relation, and describes their sequences, is applying the scientific method and is a man of science. The facts may belong to the past history of mankind, to the social statistics of our great cities, to the atmosphere of the most distant stars, to the digestive organs of a worm, or to the life of a scarcely visible bacillus. It is not the facts themselves which form science, but the method in which they are dealt with. The material of science is coextensive with the whole physical universe, not only that universe as it now exists, but with its past history and the past history of all life therein. When every fact, every present or past phenomenon of that universe, every phase of present or past life therein, has been examined, classified, and coördinated with the rest, then the mission of science will be completed. What is this but saying that the task of science can never end till man ceases to be, till history is no longer made, and development itself ceases?

It might be supposed that science has made such strides in the last two centuries, and notably in the last fifty years, that we might look forward to a day when its work would be practically accomplished. At the beginning of this century it was possible for an Alexander von Humboldt to take a survey of the entire domain of then extant science. Such a survey would be impossible for any scientist now, even if gifted with more than Humboldt's powers. Scarcely any specialist of to-day is really master of all the work which has been done in his own comparatively small field. Facts and their classification have been accumulating at such a rate that nobody seems to have leisure to recognize the relations of sub-groups to the whole. It is as if individual workers in both Europe and America were bringing their stones to one great building and piling them on and cementing them together without regard to any general plan or to their individual neighbor's work; only where some one has placed a great corner-stone, is it regarded, and the building then rises on this firmer foundation more rapidly than at other points, till it reaches a height at which it is stopped for want of side support. Yet this great structure, the proportions of which are beyond the ken of any individual man, possesses a symmetry and unity of its own, notwithstanding its haphazard mode of construction. This symmetry and unity lie in scientific method. The smallest group of facts, if properly classified and logically dealt with, will form a stone which has its proper place in the great building of knowledge, wholly independent of the individual workman who has shaped it. Even when two men work unwittingly at the same stone they will but modify and correct each other's angles. In the face of all this enormous progress of modern science, when in all civilized lands men are applying the scientific method to natural, historical, and mental facts, we have yet to admit that the goal of science is and must be infinitely distant.

For we must note that when from a sufficient if partial classification of facts a simple principle has been discovered which describes the relationship and sequences of any group, then this principle or law itself generally leads to the discovery of a still wider range of hitherto unregarded phenomena in the same or associated fields. Every great advance of science opens our eyes to facts which we had failed before to observe, and makes new demands on our powers of interpretation. This extension of the material of science into regions where our great-grandfathers could see nothing at all, or where they would have declared human knowledge impossible, is one of the most remarkable features of modern progress. Where they interpreted the motion of the planets of our own system, we discuss the chemical constitution of stars, many of which did not exist for them, for their telescopes could not reach them. Where they discovered the circulation of the blood we see the physical conflict of living poisons within the blood whose battles would have been absurdities for them. Where they found void and probably demonstrated to their own satisfaction that there was void, we conceive great systems in rapid motion capable of carrying energy through brick walls as light passes through glass. Great as the advance of scientific knowledge has been, it has not been greater than the growth of the material to be dealt with. The goal of science is clear — it is nothing short of the complete interpretation of the universe. But the goal is an ideal one — it marks the *direction* in which we move and strive, but never a stage we shall actually reach. The universe grows ever larger as we learn to understand more of our own corner of it.

REVIEW

1. How does Pearson sum up the essence of modern science?

2. What is the test which he applies to determine whether a social institution should be encouraged? Do you think he has in mind "modern business" as a social institution?

3. Why stimulate the development of the scientific method? What does Pearson mean by citizenship? Would his reasoning apply to business methods and economic dealings as well as to those which are political? Show why or why not.

4. What standards does he use to distinguish good and bad science?

5. Some one has said that "scientific method is the method of noting and classifying differences." What is meant by this statement? Does this point of view correspond to Pearson's?

6. How is the scientific method a "habit of mind"? What does Pearson mean by saying "The unity of all science consists alone in its method, not in its material"? Is there a similar unity in all business, as in "business organizations," "personnel administration," "market problems"?

7. What business and economic conflicts can you suggest, the solution of which calls for the application of scientific method?

8. Apply the point of view suggested by Pearson to such problems as the cost of living; increase in fares on street railways in view of increased operating expenses; advance in the selling price of a competitive good because of increase in cost of production; in marketing.

Why Statistics and its Methods?[1]

Probably at no other period have statistics played so large a part in our problems. We have continuous sectional enumerations. We have a decennial survey of the whole country. We make studies of feeble-mindedness, of educational possibilities, of industrial output. We are making a physical valuation of the railroads. We have taken to heart the lesson Bagehot taught in his "Lombard Street," and the Federal Reserve Board does for American banking the work that he planned for English concerns. Yet fundamental questions still go unanswered. We are content with tabulation rather than analysis. We enumerate where we should interpret. In the result, in any crisis — such as the present railroad situation — where figures are involved we have no means of interpreting at all adequately their significance.

The movement for the eight-hour day is asserted by manufacturers to be prohibitive in its cost. We have no means at our disposal of checking that assertion. Our statistics seem to have been gathered for every purpose save that of getting answers to the basic questions. We have no means of checking the relation between population and the means of subsistence. That which most painfully arrests our attempts at progress is the absence of impersonal record. The demand for the abolition of child labor was postponed for years by our fear of the formidable bill of costs presented to us by business men. We were first told that child labor was the price we had to pay for the continuance of certain industries. Then when different states passed child labor laws we were informed that the backward states could not compete with those more

[1] Taken with permission from *The New Republic*, August 26, 1916.

highly organized. We felt the wrongness of these arguments. We did not put any faith in the statistics presented for our consumption. We could only plead the virtue of experiment, and sneer into extinction the habitual conservatism of the business man. Yet all the time we dimly realized that we must pay a large price for our faith. We could not but wonder if there was not a better, a more adequate way.

As a fact, a better way exists. The devil can cite statistics for his purpose, so that, for ordinary men and women, they have been tainted with the suspicion that clings to his usage of them. The last twenty-five years have seen a revolution in statistical method. Enumeration has given way to critical analysis. Under the brilliant leadership of Professor Karl Pearson there has been evolved a new social calculus of which the first fruits even are of striking importance. We have already seen valuable results in the study of education. What, for example, is the worth of the teacher's estimate of his pupils' ability? It is clearly fundamental to have a solution to such a problem and Professor Pearson has given us a response in definitely measurable terms. Or turn to social disease. We require to know what is the actual worth of our sanatorium treatment of tuberculosis. Is the average length of life of those who are returned as cured to the general population the same as that of the normal healthy man? Theory clearly requires an affirmative answer; the result, as Professor Pearson has shown, is in fact different, so that we begin to understand that the fundamental problem is here the diathesis and that it is upon its understanding that our attention must concentrate. So, too, in the problem of wages. We require a means of interpreting the means of life in terms of every social relation that is of communal importance.

What, for example, is the cyclic relation of wage-movements to rent? What is the relation of rents to size of family? What is the relation of food prices to rent? Does a decrease in the cost of food result in a movement towards more satisfactory housing? Or take the problem of infant mortality. We are too easily satisfied with its interpretation in terms either of the mother's employment or conditions of bad environment. Modern methods of statistics enable us to go a step further. We find, for instance, that the wife works because her husband has low wages. We find that her husband has low wages because he works in a poorly paid trade entrance into which is the result either of bad physique or poor intelligence. The single problem of infant mortality is thus in fact seen to involve the whole circle of economic disharmonies. A beginning in the required direction is shown by Miss Lathrop's important reports from the Children's Bureau of the Department of Labor. Mr. Goring's great work on criminology, Miss Elderton's studies of alcoholism, Miss Barrington's on eye-sight only repeat the same results in different form.

They present conclusions from which there is no escape. The fundamental business is to measure the quality of inheritance in terms of the quality of environment. For that end we need a census survey which is not intrusted merely to competent Democrats or trustworthy Republicans. It must be a survey in which medical men, statisticians, industrial experts, educators, all obtain representation. And it must be emphasized that the old statistics are out of date. We need the application to our data of the newest instruments at our service. Interesting as our surveys like that of Pittsburgh are, they have the fundamental defect of lack of precision. The social worker who has impressions to record must record them to-day in such form as admits of statistical treatment. We have passed beyond the stage where qualitative description is possible. Here, as elsewhere, it is in quantitative expression only can we place any confidence. The ideal type-survey is the Report on the Physical and Mental Condition of Edinburgh School Children prepared by the Charity Organization Society of that city. We can measure there exactly those qualities of which we desire to know the result in social practice. What is the effect of parental alcoholism on the health of the children? How far does it affect wages? What harm does industrial instability do to the attendance of the child at school? How far does a dirty home affect the intelligence of the child? All these questions can be given a partial answer from the Edinburgh report. But we want to check Edinburgh by London and London by New York. We want a study of Chicago, of Atlantic, of St. Louis. It is upon knowledge made definite and measurable that the advance of the future will be secured.

Beginning with the great rate inquiries of 1910 Mr. Brandeis made earnest pleas for the establishment of a Bureau of Cost Accounting. We are paying the price now for our failure to take proper advantage of his counsel. Nothing would have contributed more to our understanding of the railroad situation than the ability to compare, item by item, the method and cost of operation of each railroad in the country. We could have thus obtained a kind of composite portrait of conditions which would have gone far to remove the haze and dimness of our present uncertainty. We could have known, for instance, the exact way in which the Boston and Maine Railroad has improved its earning capacity relative to the comparative failure of the New Haven Road. We would have forecasted means of improvement. We could have suggested maximum costs of output in every branch of railroad operation. If thus far we have failed in our wisdom, we may no longer wait upon the event.

We want, further, studies of wage situations such as those which Mr. R. H. Tawney is making of the industries governed by the Trade-Board Act of Great Britain. Our own students are too prone, in similar work, to describe methods of operation and give statistics of output as the right method of approach. Mr. Tawney's method is refreshingly different. He studies in definite terms the working of his industry. He explains its reaction to the wages of men and women, to prices and profits, to trade unionism. He studies in detail its effect no less on the management of industry than on the workers. He discusses the relation of minimum rates to degree and security of employment, and to home work. He makes evident the defects and virtues of the administration of minimum rates. A single, brief chapter gives us all we require to know of the actual method by which the industry is organized. The study, as a whole, is a triumphant vindication of the principles underlying the demand for the minimum wage. But it is a vindication almost uniquely valuable in industrial inquiry in that its conclusions are based on the provision of an unimpeachable bill of costs which is, from the statistical standpoint, as imaginatively conceived as it is brilliantly executed.

We in America can be satisfied with no less than this. Under the Commerce Clause trade is in the hands of Congress, the Interstate Commerce Commission, the Federal Trade Commission, both of these having their statistical departments. It is not too much to ask that the methods they apply to their problems be such as are most likely to provide the best basis for public judgment. A statistician is no longer a clerk, but a mathematician who has specialized in the theory of probability. We want men of that kind to direct our inquiries. We want all those engaged in social work to think out collectively the right questions and to analyze our material by the methods which alone give promise of sufficient response. Statistics is no longer a matter in which a single university course explaining the means by which an average is calculated is really adequate. What we need attached to every important government department and every great university is a statistical laboratory such as that of which Professor Pearson has the direction in London University. We shall then begin to know the basis upon which our social problems really rest. We shall then have satisfactory demonstration of the best lines of their efficient understanding.

REVIEW

1. Sketch the bill of complaint given expression to in "Why Statistics and Its Methods." Does this complaint have any application in the business with which you are connected or in the statistical activities of any business or agency with which you are acquainted? How?

2. Is the business man interested in these larger problems outside of "his" business? Why? What are the *limits* of *his* business?

3. Can such a problem as the eight-hour day be settled by statistics? Would statistics have any bearing on such a problem as the tariff? Why? On the establishment of a wage policy? How?

4. If statistics were collected solely to settle "basic questions" would not the occasions for collecting them be so diverse that little general knowledge would be obtained? Would such statistics meet the day-to-day needs of business men, students of economic and social conditions? Illustrate why or why not.

5. Which seems to you preferable on the part of governmental statistical agencies, (*a*) solely to collect "general purpose" statis-

ties, or (*b*) solely to collect "special purpose" statistics? Can there be a combination of both? If so, who should determine the kinds of activities to which the statistics apply? Name some general purpose statistics which are collected, which are of interest to a business man as a business man, and to him solely as a citizen. Can you think of any collected which seem primarily to answer his problems? Consult the following statistical publications with these points in mind: *The Census of Manufactures*, U. S. Census; *Monthly Summary of Commerce and Finance; Bradstreets; The Statistical Abstract*, U. S. Department of Commerce; *Iron Age*.

6. How do you explain the attitude voiced in the following statement? "We cry aloud for facts; there is a voracious and undiscriminating appetite for figures, or rather for the nourishment they afford to argument and propaganda; statesmen, teachers, preachers, publicists, and men in the street exemplify it. It is a dyspeptic appetite, if you please, because of the ill-assorted wares upon which it feeds. On the other hand, there is an almost equally common and more or less outspoken distrust of statistics or the widespread application of the statistical method as a means of obtaining working knowledge." What is to be done about this? Wherein does the difficulty lie?

7. A contrast has been drawn between what is called "statistical foresight" and "statistical hindsight." What does such a contrast mean to you? Is this distinction identical with that between "statistical planning" and "statistical planlessness?" Would statistical planning in your judgment largely correct the condition described in question 6? [1]

Statistical Control Including Costs as a Factor in Production [2]

General. — A mánager desiring to determine the best place at which to locate a particular type of retail store, considers possible locations from many points of view, in-

[1] See W. C. Mitchell, "Statistics and Government," in *Quarterly Publications of the American Statistical Association*, March, 1919, pp. 223–235.

[2] Adapted with permission from Person, Harlow S., *The Annals of the American Academy of Political and Social Science*, September, 1919, pp. 220–230.

cluding casual observations of the places where the greatest number of possible customers seem to pass. He then stations at each of these places an observer who, in a square on a tally-sheet ruled in a carefully predetermined manner, makes a mark as each person passes. After the observations have been completed and the marks in the various squares are counted, the manager is enabled to establish a number of facts pertinent to the problem such as the following: the average number of persons who pass during a day; the average who pass each hour of the day; the average number of men who pass each hour of the day; of women; of children; the number of office girls who pass during the lunch hour; etc. These group facts, discovered by recording and classifying the mass of unit facts, are of importance in helping him to decide a problem of business policy.

If a merchant sells hats for a season and keeps no record of sizes sold, he is at a loss to place precise orders for the next season. He may have a general impression that he had better place in stock more of a given size than of other sizes, but a "general impression" is not precision, control and economy in operation. On the other hand, if he has kept records, he may find he has sold 50 size $6\frac{1}{2}$; 150 size $6\frac{3}{4}$; 300 size 7; 500 size $7\frac{1}{4}$; 400 size $7\frac{1}{2}$; 150 size $7\frac{3}{4}$; etc. — in all some 1600 hats. He estimates that his sales will amount to 2000 hats next season and divides the order for that number in the ratios with respect to size, of .5, 1.5, 3, 5, 4, 1.5, etc., and feels certain that he is forecasting his market with precision.

These illustrations should suggest to the reader the nature, the purpose and the methods of statistics in business. An illustration might have been used in which facts are entered on "forms" in an office, as documents resulting from operations and carrying different kinds of data (units of product; wages; sales; complaints; prices of materials; etc.) pass through the office. The magnitude of the business, the volume of the data, the number observed, recorded, classified, compared, and otherwise handled, make no difference.

Nature and Purpose of Statistics. — A "fact," the relations of which are obscured, has little or no significance. A single person passing the observer in the first illustration has no meaning or importance. Related to the problem of locating the store he begins to assume importance. Related to that problem as one person in an aggregate of persons passing the observer, he becomes *in this relationship* of great importance; but by becoming part of an aggregate of persons he is transformed into one of a mass of data so numerous as to confuse the mind, which is limited in its processes of observing, valuing, remembering, and comparing separate experiences which come to it casually. The mind is unable to grasp the significance of larger summarizing facts behind or contained in the mass.

Yet there are summarizing facts there, facts which result from the bringing together an analysis of the aggregate. Statistics is the science and the art of handling aggregates of facts — observing, enumerating, recording, classifying, and otherwise systematically treating them — so that other "master" facts or principles or laws lying behind or contained in the aggregate are made comprehensible to the mind and become, along with the results of other methods of investigation, data for reasoning, the drawing of conclusions, the making of decisions, and the determination of policy.

Statistical Methods. — There have been developed many devices for the summarizing and analysis of statistical data such as the *per cent* and the *arithmetic average.* No manager of a plant of any size, for instance, could carry in his head the number of hirings and separations for two or three years. Yet if recorded these facts can be classified and summarized through the medium of coefficients, and the mind can easily reason in terms of the coefficients, which sum up group facts behind the unit facts. That labor turnover was 43 per cent in 1918, and 27 per cent in 1919, is the statement of two significant, comprehensible, summarizing facts yielded by proper treatment of a large number of accumulated unit facts, which considered individually had relatively little significance. The business statistician does not need, in the present stage of the development of the art of statistics in business, to go into such refinements of statistical method as are necessary for, let us say, the biologist, or even a department of public health. Extreme refinements of method yield only a fictitious accuracy when the preceding steps of observation, enumeration, and classification are lacking in precision, or the data are not in great volume, which is usually the case in a business. The accuracy of a chain of reasoning can be no greater than its weakest link.

But if refinement of method in the mathematical treatment of data is unnecessary in the use of statistics in business, too great care cannot be exercised with respect to the collection for data. The summarized data become premises for reasoning, and to the extent that they have been incorrectly labeled and classified in the process of collection and recording, the reasoning and the conclusions of which they become the basis are unreliable. The skilled statistician wants to know how the data were collected — are they complete or a good sample of the mass of unit facts under consideration; is classification exact; are compared averages the averages of like things; etc.? The critical stage in statistical investigation is the first stage; the determination of the purpose of the investigation; precise definitions of different kinds of unit facts to be recorded; the careful recording, classification, and summarizing of these unit facts in accordance with the precise definitions. From that stage on statistical processes are simple. It is in that stage that the difficulties lie and the errors are made.

Homogeneity of Statistical Units. — That the original units of observation and record should be homogeneous is the primary rule of all worth-while statistical effort. This depends upon careful definitions. If definitions are not exact, dissimilar things will be enumerated under the same head by different observers or recorders, homogeneity will not exist, and the summaries and averages will not be comparable. One recorder might include under "wages" some payments that another includes under "salaries." One might include under "worked materials" some things that another includes under "stores." One might include in the length of time it takes to perform an operation, the time between the start and finish of the operation that the machine is idle; another might not. The statistics of labor turnover published to-day are generally incomparable because of this error. In one plant "separations" is made the basis of computation; in another "hirings." In one plant the working force may be increasing, in another decreasing; neither "separations" nor "hirings" has the same significance in the one as in the other. Different unit facts are classified under the same head and the law of homogeneity is violated. Resultant averages are not comparable.

The primary statistical fact — statistical unit — observed and recorded should not be a compound fact. To use a chemical analogy, it should be an element. Compounds can be built up, if desired, by bringing elements together. The recording and analysis of homogeneous primary facts require planning ability and cost money, but they are the only facts worth recording. Later attention will be directed to the use of mechanical devices which make possible the recording and classifying of unit facts at a reasonable cost.

Statistics in Business. — The application of statistics was first developed by governments and quasi-public institutions in the study of social phenomena and was then developed and carried to the highest degree of perfection in technical method by the biologists in the study of the laws of heredity. In these fields the data have always been so numerous as to compel statistical treatment, and in these fields great discoveries have been made by the statistical method of investigation. Among business institutions the first to use statistical methods were the insurance companies, railroads, and similar businesses, the data of whose operations are voluminous and usable only when statistically handled. With the broadening of markets and the increase in the size and in the volume of business of other industrial institutions, the use of statistics increased as an aid in establishing standards, and in interpreting facts as a basis for the forecasting of tendencies and the determination of policies. To-day there are few large business institutions in the United States — manufacturing or distributive — which do not have statistical departments, and regard for the statistical function in smaller institutions is increasing with great rapidity. There is scarcely a business of any size which could not use statistics to advantage, the size of the "statistical department" being purely a problem in overhead cost to be viewed in the light of probable returns. There is an advertising company which carries an immense and costly statistical overhead, but the result of the work of that department has made the company impregnable in competition; its clients have confidence in its advice. I know of a small distributing house in which a young graduate of a school of business administration, along with other duties and on his own initiative, began to record, classify, and analyze data according to the statistical method. In one year he proved "master facts behind the mass of unit facts" never before observed, and influenced purchase policy and sales policy — for the business he effected economies resulting from operations in accordance with better policies, and, for himself, proved himself worthy to be a branch manager. Between these two extremes may be found throughout business a great variety of methods of utilizing statistics in investigation.

The Practical Objects of Statistics in Business. — The principal objectives of the use of statistics in business are:

1. To ascertain inner, controlling, master facts which cannot be ascertained by casual observation of the complex mass of obvious facts which constitute the experience of the business and in which they are contained. The sales manager about to undertake a sales campaign, does not trust to chance or to casual observation more than is necessary. He investigates and analyzes characteristics of the consuming public in a market — estimates among other things their probable demand for and capacity to purchase the particular commodity he proposes to introduce, and the kind of advertising methods to which the purchasers of that market are most likely to react. The utility corporation analyzes statistically a growing suburb, before it

determines its policy of extension and capital investment. The merchandise and credit managers of a wholesale distributing house estimate the purchasing power of a region, through the statistical analysis of crop and other governing conditions, before determining policy with respect to a season's business. The manager of a retail store may analyze sales of different articles by sizes, seasons, etc., in order to determine a quality, quantity, and seasonal schedule of purchases, thereby adjusting orders to probable turnover.

2. To determine standards by which to value and guide *current* performance and in terms of which to estimate *future* performance. The merchandise manager of a department store receives each morning a summary sheet showing sales of the preceding day compared with sales of the same day the year before; cumulative sales of the month to date compared with cumulative sales of the corresponding period of the year before and with estimates for the current month; cumulative sales of the year to date compared with those of the corresponding year before, and so on. He can ascertain at a glance whether sales are going well; if they are not he may institute at once a special sales campaign. Likewise any business selling commodities or services. A production manager time-studies operations under different conditions and with different materials and methods, and by statistical treatment of the data establishes several standards: standards of conditions; of materials; of methods; of performance. He can then value and guide current performance and can estimate with precision future performance. He may keep his record in terms of units of output and in terms of units of cost. Cost units are no different from other units in statistical treatment. A telephone company analyzes statistical records

of calls and establishes a standard of performance for an operator or for a system and on the basis of these standards can determine whether an operator is efficient or a system is approaching the volume of business for which it will be inadequate, requiring extension or replacement. The electric light or telephone or other similar company, by statistical records determines the hours, the days, and other seasons when its various peak loads are bound to occur, and establishes operating policy accordingly. A supply division of the army or navy, by statistical methods, determines a procurement and delivery schedule for an army of a given size under predetermined conditions of activity, and by similar statistical methods determines from day to day whether the schedule is being observed.

The use of statistics in determining such standards for measuring current performance and estimating future performance is one of the latest developments of the use of statistics in business, offers one of the most profitable instruments for improvement in managerial methods, and unfortunately involves some of the greatest dangers of misuse. These misuses are prevalent in current practice. The first is the error of so organizing the function of recording, classifying, and analyzing data as to secure the returns too late for use in controlling *current* operations, in which case the statistics are but records of past performance and have so limited a usefulness as to raise the question whether they are worth the cost of collection. The second error is that the units of enumeration may not be homogeneous, and to the extent that they are not, their value in the control of current practice or of forecasting future performance is invalidated. A time (statistical unit) of a performance by method A under condition B with material C on machine D is not homogeneous with a time resulting from a study when either A, B, C, or D is different. Three complaints, one resulting from disturbed mail service, one resulting from a defect in the goods, and one resulting from discourtesy of a clerk, are not homogeneous. To record them simply as "complaints" may enable a manager to enjoy the sensation that something is wrong, but will give no precise information which will enable him to control the situation and remedy the causes. The third error in the use of statistics in establishing standards and measuring performance is that the units of statistical record may not correspond to the units of the operating processes. This is a common error, for only too frequently the statistical function is not recognized as a production function, and the statistical department and methods are developed independently of the production department and methods. The analysis of processes by the production manager for the purposes of operating control is different from that of the statistical department for purposes of record, with the result that the statistics fail to be useful to the production manager. The same authority that approves the establishment of production methods should approve the establishment of statistical methods in so far as they are concerned with statistics of operation, in order to insure that the units of statistical record shall be identical with the unit process of production. Furthermore, the only way of assuring such correspondence is to make the "papers" which control production the original documents from which statistical data are drawn.

3. To establish series of facts which suggest tendencies, or permit comparisons which suggest causal relations, or at least correlation, between series. Time curves may be plotted showing sales — by salesmen, by territories, by articles, etc. By these the sales manager may keep in-

formed concerning the sales tendency in a territory, of a commodity, or of a salesman. Comparison of these curves may permit the manager to determine that the salesman whose record of gain is best is concentrating on leaders which yield small profit while a salesman whose record for gain is not so good may be selling a wider variety of articles, thereby laying the foundation of a better long-run business in his territory. Curves of wages paid, hours of work, output per man, separations, hirings, cases of discipline, idle machine time, etc., may be compared and correlations proved — *i.e.* it may be observed that when one curve shows a particular tendency another shows a similar or different particular tendency. The establishment of such correlations permits more accurate forecasting of results and the establishment of more dependable policies. There is opportunity for the development of statistics of this kind in every business and the results may be considerable, but in no two businesses is it the same, and each is a field for special study.

There are many data pertaining to the social-industrial conditions in which a business is carried on, of importance to every manager in determining policy, but to collect, classify, and analyze these would be too great a burden of cost for one business. We have in mind data relating to crop conditions, prices of basic materials of industry, bank clearings, commercial failures, etc., which when consolidated and compared throw light on general business conditions. Statistics of this sort are now available through statistical service agencies, and it is not necessary for the individual business to secure them. But there remains a considerable number of special "lines" of statistics, especially pertinent to its materials, products, and markets, which a business may profitably maintain.

4. To determine laws governing industrial operations. A comparison of different lines of statistics might disclose such relations as to prove principles to which the term "law" could properly be applied. Extraordinarily large numbers of homogeneous data are essential to the establishment of laws. These are seldom available in the records of a single industrial concern. The most noteworthy case of the scientifically precise observation, recording, classification, analysis, and general statistical treatment of industrial data which has led to the formulation of laws, was the study by Mr. Taylor and his associates which led to the discovery of the laws of metal-cutting, which revolutionized that art. The hope of the discovery of laws governing industrial operations depends upon the pooling of the statistical interests of many concerns — coöperative statistics which will yield homogeneous data in great volume.

Cost Accounting. — Cost accounting is a specialized phase of statistics. It is statistics in which the statistical units are monetary values — cents, pence, centimes. The principles of the statistical treatment of these units are no different from the principles of the treatment of other units — pounds, gallons, bushels. Cost statistics are subject to every law governing general statistics, and most of the troubles in cost accounting are the result of disregard of statistical laws. Cost statistics should be derived from operating "papers"; these papers should flow in a constant stream over the desks of cost and other statistical clerks and keep the record "up to the day" as a basis for *immediate control* of operations; the cost unit data should coincide with or dovetail into the unit data of other phases of statistics; they should be homogeneous. Costs which have been derived in accordance with these principles are worth the expense; costs which are but the record of past events — records got up too late to influence current action and in classes which do not correspond to classes of operations in the shop — are seldom worth the expense of collection.

Mechanical Devices. — The principal obstacle which is met in the development of the cost and general statistical methods here recommended is the clerical expense involved. The expense of copying data from operating forms on to special statistical department forms, and then of computations and tabulation, is frequently prohibitive. But it is possible to adapt the cards of the standard sorting and tabulating machines for use as original operation orders, and they become, after their use in operation, the data cards of the cost and other statistical clerks. One firm at least has economically secured extraordinary results in this way. The economy resulting from the use of mechanical devices and the exceptional minuteness and value of the costs and other statistics derived by this firm, are due to the fact that the statistical methods are tied up with — are a function of — the good management methods.

Graphical Records. — Graphical forms of recording statistical data — especially summarizing data — have been found desirable by all well-organized statistical departments. The simple curve is the most useful graphical device. It has properties, not characteristic of tables, which aid the mind in detecting, through the eye, tendencies and relations. There are firms which plot and keep posted daily as many as 1500 or 2000 curves.

The Statistical Department. — The statistical function should be performed by specialized clerks trained in the methods and in the manipulation of mechanical devices and in statistical operations. The manager of the department should be above all a man of imagination and of analytical ability. He should suggest, but he alone should not determine what statistics should be kept and what objectives aimed at. Statistics are for use, not for file. The executive and the administrative officers are the users. They should participate in determining what statistics should be kept. Their several desires should be dovetailed into one organic body of statistical records, coördinated by the general manager through the management engineer.

General Information: A Supplementary Function. — Statistics is a method of investigation, of securing information. It is logical therefore that other methods of securing information than the statistical should be assumed by the statistical department. Special libraries, including files of books, pamphlets, trade periodicals, and newspaper clippings, of which all the important contents bearing on the business are indexed, are being developed by statistical departments. The department should take the initiative in bringing pertinent information to the attention of the administrative and executive officers; should establish its information service within the plant.

Conclusion. — Statistical results secured in accordance with correct statistical principles and methods — related to operations, posted up to the day, based on homogeneous units — are as important to the well-managed manufacturing plant as are the sextant and the compass to the mariner. They permit the management to know at any moment *where it is* and to set its course. Statistics which yield only records of past events are of no more use than the *log* to the mariner; they do not assist one to shape one's course. Statistics are recorded, classified, and analyzed experience. From this experience, so made available, principles may be derived to guide all who are concerned with the determination and execution of policies and with the

D

direction of operations — directors and president, general manager, production and sales managers, employment manager, and others according to their respective problems. More accurate forecasting of conditions will be possible and more precise control leading to desired results; more reliable forecasts of demand, more favorable buying, better selection and training of workers and retention of workers; more precise and dependable production methods; and a better schedule of production throughout the year.

REVIEW

1. Compare the definition of statistics given by Mr. Person with that given in the *Text.* What have they in common? In what ways, if at all, are they different?

2. As in question 1, compare the definitions of *statistical methods.* Wherein do the main difficulties lie in the use of statistical methods? Is the emphasis different from, or the same as, that developed in the *Text?* How?

3. What does the author mean by *homogeneity* of statistical data? Illustrate in other fields.

4. What does the author mean by "master facts behind the mass of unit facts"?

5. Enumerate and illustrate the "Practical Objects of Statistics in Business."

6. What are the dangers of misuse of statistics in determining "standards for measuring current performance and estimating future performance"? Illustrate these further from your own experience. How are these to be overcome in statistical analysis?

7. On what types of statistics should a business concentrate its attention so far as collection is concerned, and for what types may it look to outside sources? How does your answer as a general proposition fit your own particular business problems? Illustrate.

8. How does the author define "cost statistics"? Do you agree? Why?

9. Would you say the author thinks of statistics as an end, or a means to an end? Distinguish the two points of view.

10. Does the writer's treatment support the conclusion that "statistics are recorded, classified, and analyzed experience"? Would you think it necessary in any way to expand or condition this statement? How?

Scientific Methods — The Method of Investigation in Relation to Business Cycles[1]

Beveridge ascribes crises to industrial competition, May to the disproportion between the increase in wages and in productivity, Hobson to over-saving, Aftalion to the diminishing marginal utility of an increasing supply of commodities, Bouniatian to over-capitalization, Spiethoff to over-production of industrial equipment and under-production of complementary goods, Hull to high costs of construction, Lescure to declining prospects of profits, Veblen to a discrepancy between anticipated profits and current capitalization, Sombart to the unlike rhythm of production in the organic and inorganic realms, Carver to the dissimilar price fluctuations of producers' and consumers' goods, Fisher to the slowness with which interest rates are adjusted to changes in the price level.

One seeking to understand the recurrent ebb and flow of economic activity characteristic of the present day finds these numerous explanations both suggestive and perplexing. All are plausible, but which is valid? None necessarily excludes all the others, but which is the most important? Each may account for certain phenomena; does any one account for all the phenomena? Or can these rival explanations be combined in such a fashion as to make a consistent theory which is wholly adequate?

There is slight hope of getting answers to these ques-

[1] Adapted with permission from Mitchell, Wesley C., "The Method of Investigation," in *Business Cycles*, Chapter I, Sec. III, pp. 19–20.

tions by a logical process of proving and criticizing the theories. For whatever merits of ingenuity and consistency they may possess, these theories have slight value except as they give keener insight into the phenomena of business cycles. It is by study of the facts which they purport to interpret that the theories must be tested.

But the perspective of the investigation would be distorted if we set out to test each theory in turn by collecting evidence to confirm or to refute it. For the point of interest is not the validity of any writer's views, but clear comprehension of the facts. To observe, analyze, and systematize the phenomena of prosperity, crisis, and depression is the chief task. And there is better prospect of rendering service if we attack this task directly, than if we take the roundabout way of considering the phenomena with reference to the theories.

This plan of attacking the facts directly by no means precludes free use of the results achieved by others. On the contrary, their conclusions suggest certain facts to be looked for, certain analyses to be made, certain arrangements to be tried. Indeed, the whole investigation would be crude and superficial if we did not seek help from all quarters. But the help wanted is help in making a fresh examination into the facts.

It is not feasible to make a study of all crises. . . . Not only is the field too extensive to cover thoroughly, but the recorded information is also too vague, too much confined to the dramatic events of the crises, and too scanty concerning the intervening phases of depression and prosperity. Whatever chance there may be of bettering the work already done lies in securing data more full and more precise than the data heretofore employed. The minute examination of a few business cycles therefore promises better results than a general survey of many. Hence attention will be concentrated upon those cycles concerning which the fullest and most exact knowledge is available — the cycles of the last two decades. By including England, Germany, and France, as well as the United States, a sufficient number of cases can be had to warrant generalizations.

The materials most important for such an investigation are the current reports of business periodicals and the statistical records of business activities. Most stress must be laid upon the latter; for the problems to be dealt with are largely problems of the relative importance of different factors, or of the general trend of diverse fluctuations. Quantitative analysis of the phenomena is needed quite as much as qualitative analysis. Since in his efforts to make accurate measurements the economic investigator cannot devise experiments, he must do the best he can with the cruder gauges afforded by statistics.

REVIEW

1. Business cycles occur; the explanations given for them do not agree. What is Mitchell's approach to a study of them? Does his method appear to you to be scientific? Why?

2. What are some of the "facts" to which Mitchell refers? Consult his *Business Cycles.*

3. Is there any great likelihood that there will be an agreement on all of the facts? Can the results of the facts be statistically measured? How about speculative instincts — "the willingness to take a chance"?

The Statistical Method of Discovering and Widening Markets[1]

To take the place of the old rule of thumb, catch-as-catch-can method of selling, which is gradually passing into the discard, there is appearing a real desire on the part of industrial leaders to make scientific analysis of their selling efforts. Imbued with this desire, the manufacturer (or jobber or retailer) finds that it is no such simple task to acquire the knowledge of his own business — that he formerly thought was unnecessary — but that he now believes he wants.

When the boss learns of the experience of other concerns in the development of scientific commercial methods, he begins to cast around in his own organization, trying to get information. He finds that his own manager, perhaps, is too busy to think of the questions he propounds, or that he has lived in the business so long that he can't see beyond the walls of the factory or of the office. The sales manager believes that everything is going as well as could be expected, and the boss finds that he has little sympathy with his "new fangled" ideas. He finds that the various department managers are too engrossed in the details of their own narrow fields to be of much assistance.

Sometimes the owner finds a man in his own organization who gets the right "slant" and who has the initiative and the breadth of vision to organize for collecting the information wanted. Sometimes the advertising manager is the man who fills the need. But, more commonly, if the owner is persistent enough, he looks for an infusion of new

[1] Adapted with permission from Weld, L. D. H., "A Strong Foundation for Your Advertising," in *Printers' Ink*, January 9, 1919, pp. 3–12.

blood — perhaps in the form of a new sales manager who has had experience in other fields. Sometimes, however, he decides to establish a new department, just as the manager of a manufacturing plant, when he introduces scientific management, finds it necessary to organize a separate department to make time studies and to do the planning.

Thus it has come about that in a few cases there have been established commercial research departments, whose duty it is to collect, tabulate, and interpret information about selling methods and results, and to plan methods for increasing the effectiveness of the sales organization. Sometimes this work is done fairly effectively by advertising agencies; sometimes outside organizations, or "sales engineers" are called in; but there is a growing feeling among large manufacturing and mercantile concerns that in order to get complete and substantial service, it is necessary for them to have investigating and planning departments *of their own*, and that there is a permanent place for such departments.

The larger the concern the greater the need for such a department. But what is the kind of information wanted? What are the features of sales organization and methods that are beginning to demand attention? The answers to these questions indicate in general the function of a commercial research department.

The science of commercial research has not developed sufficiently as yet, to give a very specific answer to these questions. The functions of such a department depend largely, of course, on the nature of the business, and the selling methods in use. In the case of a large business with different departments selling a variety of articles, the functions of the research department are more numerous than in the case of a smaller concern selling a single product.

The manufacturer of advertised and branded articles usually has more need of a research department than the seller of unbranded articles.

Broad Field, but Cultivation Should Be Intensive

The fundamental question which a commercial research department faces is this: How can we extend the market for our goods? But, in order to answer this question other questions have to be asked.

Are we getting the best results from our present selling activities?

What are our selling costs?

Is our distribution even throughout the country?

What share of the business are we getting?

Are the salesmen properly trained?

Are they paid in the best manner?

How often do they report, and *what* do they report?

How thoroughly are salesmen's reports analyzed?

How well do salesmen cover their territories, and are these territories laid out scientifically?

Could business in certain sections be developed by establishing branch houses carrying stocks of goods?

Then there are other questions concerning sales policies and price policies. Are prices maintained by dealers?

Are exclusive dealers used?

Are quantity prices allowed, and, if so, are they adjusted properly?

How do dealers feel toward our products?

Are dealers sold in proper quantities?

How many different competing brands do dealers handle?

To what extent do consumers ask for our product by its brand name?

And then there are numerous questions to be asked concerning the advertising.

These are only a few of the questions that a commercial research department might be called on to answer. It is not necessary, however, for such a department to start out by trying to solve all the problems suggested above. Rather may it prove more useful by addressing itself to some specific problem.

Perhaps the most important service that a commercial research department can perform is the collection of information that can be obtained only by field analyses or market surveys — that is, information that does not exist within the organization in any form, but that has to be gathered from the outside. The only members of the organization who could possibly have this information, or who are coming in contact with the people from whom it could be obtained, are the salesmen.

But salesmen can't successfully make the market surveys necessary in scientific selling for the following reasons: (1) If a salesman is properly routed over his territory, he cannot possibly have the time to collect the information needed; (2) The salesman has a personal interest, which blinds him, either consciously or unconsciously, to facts that would place his work in an unfavorable light; and (3) many salesmen are lacking in a broad conception of fundamental merchandising problems, and hence they frequently fail to grasp the significance of facts which would be of value to the management.

For these reasons, market surveys need to be made by men who are detached from the regular selling force. Furthermore, they ought to have a training in the fundamentals of business organization. They ought to be able to answer intelligently: Why does my firm sell through jobbers, rather than direct to retailers? What would be the advantages of selling direct? How much more would it cost? How much, approximately, does it cost to sell the different commodities my concern is marketing, and what is the relative profitableness of the different lines?

Questions Swift and Company Are Solving

A good example of the difficulties surrounding this last question is a problem faced by Swift and Company. This company sells a variety of products through its 400 branch houses. Branch-house selling costs are measured as so many cents per hundred pounds — lumping together "Premium" hams, oxtails, soap powder, eggs, oleomargarine, etc. Just how much it costs to sell soap powder as compared with "Premium" hams, can never be determined exactly, but approximations can be made by considering amount of salesman's time necessary to sell, rate of turnover, amount of storage space required, etc.

This suggests another of Swift and Company's selling problems. Goods are distributed partly through branch houses and partly by means of "car routes." Car route distribution means the supplying of retailers in small towns direct by drop shipment from refrigerator cars that are sent out from the packing plants at regular intervals, each car serving the dealers in a dozen or more towns along a line of railroad.

This question frequently arises: Shall a certain town be served by a car route or shall it be served by a near-by branch house, or is the town large enough to have a branch house of its own? Only when one gets beneath the surface, can he begin to realize the complexities of this problem, especially when a perfectly commendable business rivalry and jealousy between the two departments involved has precluded the development of a scientific method of answering this question when it arises. This is only one of many instances that suggest the possibilities of a commercial research department for such a large concern as Swift and Company.

Even if market analyses and surveys are the main object of a commercial research department there are certain statistical analyses of existing facts and figures which should be made first. There are very few concerns that have analyzed to their fullest possible usefulness, the figures that they already have in their own records. Many firms have, within the past few years, forced their salesmen to go to the trouble of making daily instead of weekly reports, and then have not themselves taken the trouble to make proper use of the information furnished by such reports. . . .

Analysis of either existing facts, or of facts that have to be obtained by means of field surveys, calls for a knowledge of statistical methods. The construction of averages, of per capita sales by states, etc., offers many pitfalls to the uninitiated. Improper statistical analysis may do more harm than good. . . .

Use of Graphic Charts

One of the most valuable things that a research department can do in a statistical way is to present its analyses in the form of graphic charts. Curves representing sales by weeks or months are invaluable. The writer believes that the common practice of comparing "last week's sales" with the sales of the "corresponding week previous year," is hardly sufficient to give an accurate picture of sales development. The sales of the different products should also be graphed. The seasonal variations should be studied — and for different sections of the country. Then these things should be compared with the methods of routing salesmen, the possible effect of changes in advertising policy, etc. "Graphic control" of industry is becoming recognized more and more. It saves the time of executives, and it gives them a broader view of their business problems. . . .

Market surveys may cover either dealers or consumers, or both. Consumer surveys are necessarily the more costly, in that they require more time and a larger corps of investigators. Much information about consumers may, of course, be obtained from dealers.

It is, of course, not necessary to visit all retailers or all consumers in the country! The method of "sampling" may be used. Typical communities in different parts of the country should be carefully selected. After the returns have begun to come in and are tabulated, it is possible for the analyst to determine how comprehensive the survey must be in order to make it yield accurate and dependable results. When the returns from different communities begin to check with each other and show the same tendencies or explainable differences, this is an indication that dependable results are being obtained. When they show irreconcilable and unexplainable differences, this is an indication that the survey is not comprehensive enough to bring forth trustworthy fundamentals.

In planning a survey, a list of questions should be drawn up as carefully as possible, worded in such a way as to be answerable in the easiest possible way. Whenever possible, questions should be asked in such a way as to be answered by "Yes" or "No" or by some figure. A list of questions should be tried out before the final form is adopted. The man in charge of the investigation should do some of the field work himself, in order to be able better to interpret the results, and to understand the difficulties of the investigators. The question should be printed on forms of convenient size and shape and on good enough paper to be easily handled and read. . . .

From dealers, the manufacturer wants to know how many lines of competing goods are carried; what percentage of the business he is getting; whether consumers ask for the article by its name; whether dealers push certain brands, and why; why goods are returned; whether store signs and dealer helps are used, etc., etc.

From the consumer the manufacturer wants to know why she buys, or why she doesn't buy, his product; whether retailers try to get her to buy a substitute; whether she likes the color and the appearance; how often she buys, or why she doesn't buy, his brand, etc., etc.

This is the kind of information that can be obtained in the best possible way only by a commercial research department. There are also many problems in connection with advertising methods and copy that can be solved only by personal contact with dealers and consumers; and it should be the duty of this department to help in the analysis of advertising results, and to check up the agency on the choice of mediums, etc.

From the foregoing analysis it would seem that there are enough things for a commercial research department to do, and there is probably not a single one in existence that has tackled half the things enumerated. The usual experience has been, so far as the writer knows, that such a department has found itself so busy with just a few specific problems, that it has proved its usefulness even within restricted fields, and has unbounded possibilities ahead.

In conclusion let it be said that in many industries there are still other problems than those mentioned above, to which a research department may well address itself. And these are some of the most vital problems of the day. These have to do with the broad and fundamental relations of an industry with the public and with the Government. The economics of any industry are well worth studying. Just what economic function does any particular industry perform? How is it a benefit to mankind? To what extent is it misunderstood by the public? How can its service be improved? What is its policy in dealing with the public and with its own working people? . . .

REVIEW

1. Has the author a scientific viewpoint respecting advertising and market extension? If so, why? If not, why not?

2. What is meant by commercial research? In studying markets, what kinds of questions must a commercial research department ask in order to be "scientific"? Where and through whom are answers to such questions secured?

3. In what way is Swift and Company in need of a commercial research department? Would your answer apply equally well to all types of businesses? Can you defend the establishment of a research department in a country bank, in a grocery business? Is the size of the department alone significant in the application of "scientific method"?

4. Can "guesses," to which exception is taken, be scientific?

5. To whom should market surveys extend? Compare the discussion of "sampling" as here treated with the discussion in the *Text*.

CHAPTER II

SOURCES AND COLLECTION OF STATISTICAL DATA

Statistics of Unemployment [1]

Statistical information as to unemployment in the United States is less adequate and reliable than that as to almost any other social problem. The federal government, several of the states, and various other agencies have made censuses of the unemployed from time to time, but in the greater number of cases the data thus secured are of little value. . . .

The sources of statistical information as to unemployment among trade unionists are the publications of the state departments of labor and of the trade unions. . . .

The New York Department of Labor has collected since March, 1897, statistics of unemployment among the trade unionists of that state. From 1897 to 1914 it collected semi-annually, from all the trade unions, information as to the number of members employed and unemployed on the last working days of March and September, the causes of such unemployment, the number of members idle throughout the first and third quarters of the year, and the number of days which each member worked during these periods. The supply of this information was made compulsory by law. Since December, 1901, the New York Department has selected certain local unions in each trade

[1] Adapted with permission from Smelser, D. P., "Unemployment and American Trade Unions," *Johns Hopkins University Studies*, Series XXXVII, No. 1, 1919, pp. 9–32.

and industry from which it has secured monthly returns as to unemployment. It has attempted to select local unions which have reliable and intelligent secretaries, to have each trade represented in proportion to the number of workmen engaged in each class, and to maintain the same proportionate representation from month to month so that the data may be comparable.

Both classes of statistics are of doubtful value. The secretaries of the local unions in many cases had no means by which they could determine the actual number employed and unemployed, and consequently they resorted to rough estimates. Further, there was a tendency to exaggerate the amount of unemployment in the hope that this would favorably affect public opinion. These defects were especially inherent in the data collected semi-annually from all unions, and for this reason the collection of this class of data was discontinued in 1914. The data relating to selected unions are defective in many respects, but it is thought that, while they are of no great value as regards the actual amount of unemployment, they are of considerable importance in making apparent the movements in the state of employment from month to month and from year to year. . . .

The Massachusetts Bureau of Statistics, since March, 1908, has collected data as to unemployment from trade unions situated in that state. This information is comparable, in many respects, to that collected by the New York Department. In Massachusetts information as to unemployment is secured only from those unions which desire to report their working conditions. However, the majority of the trade-union membership is represented in the returns. Thus, for the quarter ending September 30, 1915, returns were made by 1052 local unions representing 175,754 organized wage earners, or approximately 75 per cent of the trade-union membership of the State. Monthly returns are not made by any of the unions, reports being made only for the last working days of the four quarters of the year by the secretaries of the local unions. The returns are scrutinized by the bureau's experts and if any errors are apparent the schedules are returned for correction. . . .

The New Hampshire Bureau of Labor is the only other state bureau which has collected statistics of unemployment among organized wage earners, and these statistics are practically valueless as they give only the percentages of members unemployed throughout the first and second quarters of 1915. It seems that the secretaries of the local unions, in most cases, were unable to accurately report such information.

A number of the American trade unions have attempted to collect statistics of unemployment of their members. Generally these attempts have failed, either because the secretaries of the local unions refused to report conditions accurately, or because the secretary of the national union failed to recognize the importance of the statistical information as to unemployment. The unions have the opportunity of collecting such material at small expense. In all unions the secretaries of the subordinate branches make monthly reports to headquarters concerning various subjects, and where statistical information as to unemployment has been collected these monthly reports have generally been utilized for this purpose.

The American Federation of Labor collected from 1899 to 1908 data relating to unemployment among members of its affiliated unions. The number of workmen represented in the returns varied as much as 800 per cent from one month to another in the same year, and as the reports were made by the secretaries of the national unions it is obvious that the data secured were not accurate. For this reason the collection of this information was discontinued in 1909.

The Wisconsin State Federation of Labor has collected statistics of unemployment from its affiliated unions since 1912. The information collected in 1912 was worthless and that for the two succeeding years was far from satisfactory. In 1913 the affiliated unions were requested to report the percentages of members unemployed on September 1. Returns were made by 243 local unions with a total membership of 19,921. Of these, 1436 members, or 7.2 per cent, were reported as idle. This percentage is but four tenths of one per cent higher than that of Massachusetts for September 30 of the same year, while it is 12.8 lower than the New York percentage for August 31.

A few unions have realized the benefits accruing from the collection of statistical information as to unemployment and have accordingly provided in their constitutions that the local union secretaries shall report the state of employment at specified periods. For example, the Potters, Plumbers, Boilermakers, Iron Molders, Lithographers, Elevator Constructors, and Metal Polishers require the secretaries of their subordinate unions to report either monthly or quarterly the number of members employed and unemployed. But little attention is paid by the secretaries to these provisions, and in the unions where the information is reported it is neither used by the general secretaries nor compiled for publication.

The Painters, Paperhangers, and Decorators, at their convention in 1913, provided that an official "time book" should be issued to each member of the union, who was to record in it all time lost through unemployment and the causes of such idleness, and report quarterly to his local union. The secretaries of the subordinate branches were instructed to compile these reports and send them to the national union. It was thought that much valuable information could thus be secured. Considerable light would have been thrown upon the question of variation in unemployment among localities. However, it was found impossible to secure the desired information from the members except through a system of fines, which, of course, would have had a tendency to produce inaccurate statistics. Consequently, these time books are used in only a few unions. It is understood that the Chicago local union has collected statistics of unemployment from its members for five or six years. It was reported at the convention in 1913 that the data collected in the two previous years indicated that the average painter lost ninety-eight working days each year through inability to secure work.

The Glass Bottle Blowers have collected and privately published statistical information as to unemployment among its members for several years. But in consequence of the fact that no distinction is made between the members totally unemployed and those working as "spare men" this information is of little value. There is also available in the monthly journals of the Wood Carvers data as to the number of members employed and unemployed on the last working day of the month. Percentages of unemployment have been calculated for the period 1909–1915, and there is little fluctuation in them from month to month and from year to year, the rate of unemployment ranging between twenty and twenty-five per cent. This would seem to indicate that the returns are not accurate but mere estimates of the secretaries. . . .

In view of the fact that so little attention has been given to the collection of data as to unemployment in the United States before 1900, it is rather surprising to find that the Bricklayers' Union, organized in 1865, collected semi-annually statistics of unemployment from 1882 to 1911 and monthly thereafter. These statistics are based upon the reports by the local secretaries of the number of members employed and unemployed. Not all of the unions reported, as some were always in a state of disorganization or were involved in labor disputes; but the reports are fairly representative of the entire membership, and the average percentage of the membership included in the data for the period 1882–1911 is 79.1. There is no reason to believe that those unions which are not represented in the returns, except the few on strike, had more or less unemployment than the average of those reporting. The returns unfortunately include members who were reported as unemployed on account of labor disputes and illness. Of course the inclusion of these members has produced high percentages of unemployment.

Another important question is whether the secretaries correctly reported the number of the unemployed. Secretaries of unions having less than fifty members could easily determine the number of unemployed, since they generally knew the places where members were at work; but in unions with a larger membership — many of the local unions have from 100 to 7000 members — the secretaries were unable to make exact returns from their own knowledge. In such cases the secretaries either based their returns upon rough estimates or upon the reports of the stewards. It is impossible to determine the extent to which the stewards' reports were used. It would not have been difficult to ascertain the exact number of members employed on a given day if these reports had been used, because each week the stewards on the various jobs reported the names of all members working on particular days. The reports are supposed to give the number of members employed and unemployed on the last working days of June and December; but it is understood that frequently the returns were based upon the condition of trade slightly before and after these dates. . . .

The Flint Glass Workers have collected quarterly statistics of unemployment since 1907, but the data are fragmentary from 1907 to 1912. In 1913 the union also included in its inquiry questions as to the number of members who were unemployed at the trade, but who had secured temporary employment in other lines of industry. Accordingly, the local unions were requested to report the number of members employed at the trade, the number holding honorary membership, disabled, and working outside the trade, and the number of those who were willing and able to work but had not found employment of any kind.

The fact that many workmen secure subsidiary employment when they are unable to secure employment at their principal occupations is a factor that has frequently been overlooked in discussions of unemployment statistics. The fact that the unions in a particular trade report that 30 per cent of their members were unemployed on a certain day should not be construed to indicate that 30 per cent of their members were not working, but that 30 per cent were not engaged at their principal occupation. This defect in trade-union statistics of unemployment is due to the fact that the secretary of a local union estimates the percentages of unemployment with the idea that the information which is most desirable is that relating to the number of members who are unable to secure employment under the jurisdiction of the union.

Statistical information as to unemployment among the members of the Pattern Makers' Union is available for each month since April, 1907. These data have been secured from the reports of the local union secretaries to the national president who compiles the statistics for private use and for publication. The secretaries are instructed to "give the exact number of members unemployed at the end of the month" and the membership of the local unions. These statistics are, of course, open to the same criticism as those of the New York Department of Labor and Massachusetts Bureau of Labor, but they are greatly superior to the statistics collected by trade unions that have heretofore been considered. In January, 1915, forty of the sixty-five local unions of the Pattern Makers had less than fifty members each. As was stated above, the secretaries of local unions with few members are able to determine the number of unemployed from personal knowledge. Moreover, several of the larger unions, two of which comprise over 20 per cent of the entire membership, pay out-of-work benefits, and all of the local unions furnish out-of-work stamps free to the unemployed, so that their secretaries, unlike those of most unions, have the opportunity of ascertaining the exact number of unemployed members with but little difficulty. The president of the union, too, takes great interest in the returns and where a local union attempts to conceal a good condition of trade by the return of an exaggerated number of unemployed, does not hesitate to correct the error. However, President Wilson states that, although the greater number of unions make fairly accurate returns, some associations overestimate the number of unemployed for the purpose of deterring the traveling members from transferring to them. Thus, in January, 1915, he pointed out that "one association this month reports that 20 per cent of its members are out of work while the truth is that all of its members are employed, and another union reports just about three times as many as are really idle." As with the other data as to unemployment in trade unions, these figures include those unemployed from all causes. . . .

One of the most important conclusions to be drawn from the statistics of unemployment relates to the very great differences in the amount of unemployment among localities. The dominant industries of any two States are rarely the same, or even if the same, the proportions of workmen employed in the various industries are generally different. It is certainly true, for example, that the chief occupations of the workmen included in the Massachusetts returns are not identical with those of the workmen represented in the New York data. Even where the industries are the same in two States certain local peculiarities may affect the seasonal fluctuations and produce more unemployment in one State than in another. . . .

Not only are the fluctuations in employment in the industries of two States taken as a whole often quite different, but it frequently happens that the seasonal fluctuations in the same industry are different in two States. This arises chiefly out of climatic conditions although various local peculiarities play a large part. Thus, when the state of employment in the building trades of New York City is poor, Philadelphia may be erecting a number of large buildings and may need additional workmen. Indeed it may be said that the state of employment in certain trades is affected more by purely local variations than by seasonal and cyclical fluctuations. It will occasionally happen that in a particular city more building will be done during the winter than was done in the preceding summer. Even taking the labor market as a whole, the state of employment varies as much from one city to another as it does from one season to another. This fact is shown by the reports of the Massachusetts Bureau of Statistics on the state of employment in the various cities of the State. In March, 1915, for example, the percentage of unemployment for the entire State was 16.6; in Boston, it was 13.9; in Brockton, 27.6; in Holyoke, 25.2; in Lowell, 7.4; while in Quincy and Taunton it was only 4.1 and 4.7, respectively. Thus, there was a total range of 23.5 from one city to another in the same State. The reports of the New York Department of Labor show that the state of employment is generally far worse in New York City than in other parts of the State. . . .

The most noticeable characteristic of the statistics is the wide fluctuation in the percentages of unemployment from month to month. In the New York data, which constitutes the only statistical information as to unemployment from month to month in all trades, the percentages for all trades taken together gradually dropped from January, the dullest month in the year, to September and October, and rose again in November and December. The good and bad seasons vary from one trade to another. Thus, the winter months furnish less employment in building trades and transportation, but more employment in clothing, textiles, boots and shoes, theaters and music. The differences among the various trades of the same industry are equally as important. For instance, in the garment industry, the dull seasons in dresses and waists coincide with the periods of fairly intense activity in the manufacture of petticoats. While the seasons of activity and dullness may be in general the same in some of the various industries, the duration and the intensity of the unemployment may be different. In the clothing industry the seasonal fluctuations are the greatest, for in some of its trades there is an almost complete stagnation in the dull season. On the average, it may be said that the dull season affects 80 per cent of the workmen in the clothing industry. In the building trades the fluctuations due to weather conditions mean the idleness of 20 per cent of the workmen in addition to the number normally idle. In metals and machinery and printing, the seasonal fluctuations are less, amounting to but three or four per cent of the workmen. In the brewing industry the seasonal fluctuations mean the employment of all workers on half time, while in theaters about 75 per cent of the workmen are unemployed during the summer months. . . .

It is a well-recognized fact that wages are higher in trades which are affected by pronounced seasonal fluctuations than in trades embracing the same class of workmen but with greater regularity of employment. Thus, the hourly wages of bricklayers are considerably higher than the wages of carpenters; but the statistics of the New York Department of Labor show that the average yearly earnings in the two trades are about the same. Cabinet makers receive lower wages than carpenters partly, if not entirely, because they have more regular employment. The relatively high daily wages of members of building-trades unions are frequently used to indicate high yearly earnings, yet it is found that the latter are but little more than those in metals and machinery and slightly lower than in printing, where regular employment produces high yearly earnings although the daily wage is relatively low.

REVIEW

1. In what sense or senses is the word "unemployment" used by the author? By the different collecting agencies?

2. What are the sources of statistics on unemployment in the United States according to the author?

3. What general criticisms from a statistical point of view are applicable to unemployment statistics collected from trade-union sources in New York and Massachusetts?

4. What statistical success have American trade unions had in collecting unemployment statistics concerning their own members? Has this been generally true? To what fundamental condition is this due? Can the conditions be changed in your judgment? How?

5. From the unemployment data extant what are the most important conclusions which may be drawn? Would you think these statistically significant in view of the nature of the returns?

6. Do the major fluctuations in employment result from seasonal or geographic influences? Can your answer be general? Why?

7. What supplementary light does the rate of wages throw on unemployment?

8. Contrast "unemployment" and "fluctuations" in employment. From what points of view might they be used as equivalent in meaning? When would it be necessary to discriminate between them?

9. Consult the most recent of any of the Reports on Unemployment to which reference is made by Dr. Smelser. Are the data given in tabular or graphic form?

(1) Is unemployment, as used, defined; are the conditions to which the data refer clearly indicated; are you in doubt in any way respecting the significance of the data either absolutely or comparatively? In what respects?

(2) To whom would the data seem to be of interest? For whom were they prepared?

A LIST OF SERIES NOW AVAILABLE SHOWING VOLUME OF PRODUCTION IN THE UNITED STATES[1]

Agricultural Products:	Unit	Description of Series	Period	Available Since[2]	Issued By
Foodstuffs					
Butter	Pound	Production Storage	Month	1917	Bureau of Markets
Cheese	Pound	Production Storage	Month	1917	Bureau of Markets
Eggs	Case	Storage	Month	1917	Bureau of Markets
Oleomargarine	Pound	Production Storage	Month	1917	Bureau of Markets
Cattle	Number	Number Slaughtered	Month	1917	Bureau of Animal Industry
Calves	Number	Number Slaughtered	Month	1917	Bureau of Animal Industry
Swine	Number	Number Slaughtered	Month	1917	Bureau of Animal Industry
Beef	Pound	Dressed Meat	Month	1918	Bureau of Animal Industry
Veal	Pound	Dressed Meat	Month	1918	Bureau of Animal Industry
Pork	Pound	Dressed Meat	Month	1918	Bureau of Animal Industry
Pork	Pound	Storage Frozen, Salt	Month	1917	Bureau of Markets
Lard	Pound	Storage	Month	1917	Bureau of Markets
Wheat	Bushel	Production	Year	1868	Statistical Abstract
Wheat Flour	Barrel	Production, Stocks	Month	1910	Dept. of Agriculture
Corn	Bushel	Production	Year	1868	Statistical Abstract
Oats	Bushel	Production	Year	1868	Statistical Abstract
Rice	Bushel	Production	Year	1868	Statistical Abstract
Barley	Bushel	Production	Year	1868	Statistical Abstract
Buckwheat	Bushel	Production	Year	1868	Statistical Abstract
Potatoes	Bushel	Production	Year	1868	Statistical Abstract
Sugar	Pound	Production	Year	1868	Statistical Abstract
Lumber					
Lumber	Feet	Production, Shipment	Week		Nat. Lumber Mfrs. Assn.
New Business	Feet	Orders	Week		Nat. Lumber Mfrs. Assn.
Textiles					
Raw Cotton	Bales	Production	Year	1899	Census, Statistical Abstract
Raw Cotton	Bales	Consumption	Month	1917	Bureau of the Census
Spindles	Number	Number Active	Month	1917	Bureau of the Census
Wool	Pound	Consumption	Month	1918	Bureau of Markets
Active Machinery	Number	Number of Spindles, Width of Loom	Month	1918	Bureau of the Census

[1] Adapted with permission from "A Continuous Census of Production: A Round Table Discussion," *Quarterly Publications of the American Statistical Association*, March, 1920, pp. 68–69.

[2] In the case of certain series, it may be possible to secure figures for an earlier date than that here stated.

MINERAL PRODUCTS:	UNIT	DESCRIPTION OF SERIES	PERIOD	AVAILABLE SINCE[1]	ISSUED BY
Coal					
Bituminous	Tons	Production	Week	1917	U. S. Geological Survey
Anthracite	Tons	Production	Week	1917	Anthracite Bureau of Information
Coke	Tons	Production	Week	1917	U. S. Geological Survey
Copper	Tons	Production, Stocks	Week	1917	U. S. Geological Survey
Iron and Steel					
Pig Iron	Tons	Production	Month	1883	American Iron and Steel Institute
Steel Ingots	Tons	Production	Month	1901	American Iron and Steel Institute
U. S. S. C. Orders	Tons	Unfilled Orders	Month	1910	U. S. Steel Corporation
Petroleum					
Crude Petroleum	Barrels	Production, Stocks	Month	1917	Bureau of Mines
Gasoline	Barrels	Total Output, Stocks	Month	1918	Bureau of Mines
Kerosene	Barrels	Total Output, Stocks	Month	1918	Bureau of Mines
Lubricating, etc.	Barrels	Total Output, Stocks	Month	1918	Bureau of Mines
MANUFACTURED PRODUCTS: (Not listed above)					
Leather					
Hides and Skins	Number	Raw Stocks	Month	1919	Bureau of Markets
Boots and Shoes	Number	Production	3 Mos.	1918	Manufacturers Assn.
Stocks	Pounds	Stocks on Hand	3 Mos.	1918	Manufacturers Assn.
Paper					
Pulp	Tons	Production	Month	1918	Fed. Trade Commission
Paper	Tons	Production by Grades	Month	1918	Fed. Trade Commission
Wall Paper	Tons	Production	Month	1918	Fed. Trade Commission
Felts and Building	Tons	Production	Month	1918	Fed. Trade Commission
Railroad Supplies					
Cars	Number	Output	Month	1918	Ry. Car Mfrs. Assn.
Locomotives	Number	Output	Month	1918	U. S. R. R. Administration

[1] In the case of certain series it may be possible to secure figures for an earlier date than that here stated.

Manufactured Products:	Unit	Description of Series	Period	Available Since[1]	Issued By
Shipbuilding					
Ships	Tons	Production, Schedule	Month	1917	U. S. Shipping Board
Ships	Number	Officially Numbered	Month		Bureau of Navigation
Tobacco					
Cigars	Number	Production (large and small)	Month		Comm. of Internal Revenue
Cigarettes	Number	Production	Month		Comm. of Internal Revenue
Chewing, Smoking	Pounds	Production	Month		Comm. of Internal Revenue
Railroads:					
Average miles of road operated			Month	1917	
Net ton miles			Month	1917	
Net ton miles per mile per day			Month	1917	
Train miles			Month	1917	U. S. Railroad Admin. Summary O.S. 5
Net ton miles per train mile		By Companies and Groups with Totals for U. S.	Month	1917	
Freight car miles (loaded, empty)			Month	1917	
Net ton miles per car per day			Month	1917	
Gross ton miles			Month	1917	Summary O.S. 7
Passengers carried one mile			Month	1917	Summary O.S 8a.

[1] In the case of certain series it may be possible to secure figures for an earlier date than that here stated.

REVIEW

1. Consult any one of the series listed above, and determine, if possible,

(1) the definition of the unit used.
(2) the source of the data published.
(3) the method by which the data are collected.
(4) the nature of the critical comments supplied with the data.
(5) the apparent purpose which the data are to serve.
(6) the consumer to which they are addressed.

2. In what way, if at all, could these series of data be of use, for

(1) planning internal manufacturing problems?
(2) measuring market trends?
(3) measuring industrial growth?
(4) helping to indicate or solve employee-employer relations?

Sampling of Coal [1]

The standard specifications require that a sample of each delivery of over twenty-five tons of coal be analyzed to determine its quality and the acceptability of the shipment. The important feature of sampling is to secure a quantity representative of the coal delivered.

Sampling in the field at the point of delivery shall be done under the control and supervision of the borough engineer of the borough within whose limits the delivery is to be made. When the sample is taken from a pile, boat, or car, care must be taken to secure it from various parts and in the same amounts from the top, the middle, and the bottom. When coal is unloaded by conveyor, samples shall be taken by hand or mechanical means from the moving mass at regular intervals.

[1] Adapted with permission from *Bulletin No. 2, Bureau of Economy and Efficiency. City of New York, Department of Water Supply, Gas and Electricity*, pp. 27–29.

The gross sample must contain the same proportion of lump and fine coal as exists in the whole shipment. In order to avoid gain or loss in moisture samples are protected from the weather by being placed in a covered receptacle until the gross sample can be quartered down and sent to the laboratory. The size of the gross samples to be taken depends upon the size of the delivery. The standard specifications provide that:

For deliveries over 25 tons and less than 100 tons the gross sample is 200 pounds.[1]

For deliveries over 100 tons the sample is approximately one ton in each thousand tons (except where otherwise specifically provided).[1]

If the sample of coal is larger than the pea size it is broken down by hand or by passing through a crusher to approximate the size of pea (which passes through a $\frac{3}{4}$-inch square mesh and over a $\frac{1}{2}$-inch square mesh).

After being reduced to a standard size the gross sample shall be thoroughly mixed by shoveling it over and over, and is then formed into a conical pile by shoveling the coal from the edges. When the cone is completed it shall be cut in half vertically by passing a piece of sheet-iron down through the center and see-sawing it until it strikes the floor. The two halves shall then be separated by holding the iron plate firmly vertical and moving either half of the cone about one foot away. The iron plate shall then be set at right angles to its first position and the cone divided into quarters. Two diagonally opposite quarters are rejected. In the two remaining quarters the larger lumps are broken down to $\frac{1}{2}$ inch or smaller. The two quarters are then thoroughly mixed, formed into a conical pile, and quartered as before. The operation of break-

[1] Sample of shipment less than 25 tons shall not be taken.

ing down, mixing, and quartering is to be continued until the sample has been reduced to about 5 pounds and to $\frac{1}{4}$-inch size or smaller.

The sample shall be worked down as rapidly as possible to avoid change in the percentage of moisture through exposure to the air.

REVIEW

1. In what respects does the analogy between sampling as a process in coal selection and sampling as a statistical device for characterizing a labor force, for instance, seem to be complete? In what ways imperfect? Would you purchase labor on the basis of samples? Is it done?

2. Generalizing on your answers to the question above, formulate in writing a general statement of the conditions to be observed in statistical sampling.

3. What would you say, from the point of view of sampling, about figures purporting to show the average depth of spring and fall plowing in a number of States, or the statement that "in Illinois fall plowing is deeper than spring plowing, whereas in Indiana, the reverse is true. . . ."? [1]

Government Crop Reports [2]

The practical value of the Government crop estimates results from the fact that they are based upon reports of farmers and others in every county and township in the United States and upon reports of trained field agents in each State; they are made monthly during the crop season; they are checked up from every possible source of information; the final reports are prepared and issued by a crop-reporting board of experts; and all Government

[1] *Monthly Crop Report*, February, 1918, p. 17.

[2] Adapted with permission from "Government Crop Reports: their Value, Scope, and Preparation," *United States Department of Agriculture, Bureau of Crop Estimates, Circular 17*. Revised, pp. 8–26.

employees engaged in the preparation of the crop estimates are prohibited by law from giving out information concerning them or in utilizing information so obtained for their own benefit directly or indirectly prior to the date and hour of publication, so that the reports when issued are known to be as accurate as it is practicable to make them, as well as impartial, disinterested, and therefore dependable. No public organization, and certainly no private corporation in the United States and probably in the world, is so well organized and equipped for the work of reporting on crop conditions and prospects as the present Bureau of Crop Estimates.

Without such a system of Government crop estimates, speculators interested in raising or lowering prices of farm products would issue so many conflicting and misleading reports that it would be practically impossible for any one, without great expense, to form an accurate estimate of crop conditions and prospects. Farmers would suffer most from such conditions, because they are not so well organized as other lines of business nor are they in a position to take advantage of fluctuations in market prices.

Farmers are benefited by the Government crop reports both directly and indirectly; directly, by being kept informed of crop prospects and prices outside of their own immediate districts, and indirectly, because the disinterested reports of the Government tend to prevent the circulation of false or misleading reports by speculators who are interested in controlling or manipulating prices.

The farmer cannot, by refusing to report the condition of crops for his locality, prevent buyers and speculators from knowing the condition of the crop. It is well known that speculators and large dealers in farm products do not depend entirely upon Government reports for information concerning crop prospects. They maintain regular systems of their own for collecting crop information. They have traveling agents and correspondents (usually local buyers) throughout the United States who keep them posted, and the large buyer or speculator, in return, gives these local buyers or correspondents information in regard to general conditions and prices. The local buyers know the conditions of crops in their own vicinity better, as a rule, than the average farmer, because it is their business to keep well informed.

If the Government crop estimates should be discontinued, the farmer would have no reliable information concerning crop prospects except in his own immediate neighborhood, and for crop prospects in other localities he would have to depend upon such information as interested spectators and dealers might choose to publish in the newspapers, which might or might not be correct. Prices in his own local market are influenced, as a rule, more by the condition of the whole crop throughout the State or the United States, and even in foreign countries, than they are by local conditions. The entire wheat crop of his county may be destroyed and yet prices may be low, or his county may have a bumper crop and prices be unusually high, depending upon whether or not there is a surplus or deficiency in the entire crop elsewhere. In a sense the Bureau of Crop Estimates is a form of farmers' coöperation, wherein each farm crop reporter gives information about his locality and in return receives information about the entire country, the bureau merely acting as a clearing house for such coöperative exchange.

Some of the private crop reports which are published in the newspapers are honestly prepared and are more or less accurate, depending upon the extent and sources of information; on the other hand, misleading crop reports are known to be frequently circulated in order to raise or lower prices in the interest of speculators. If the farmer reads the crop estimates and forecasts of the Government as they are issued, he will be in a position to judge for himself what the crop prospects are, as well as probable prices, so that he can decide intelligently how to market his produce and how to deal with the local buyers. Even the farmers who do not keep posted are indirectly benefited by the publication of Government crop estimates, because these estimates automatically tend to check and lessen the injurious effects of false reports sent out broadcast by interested speculators and their agents in the same way that a police or constable force tends to check but not entirely prevent crime in a community.

The more certainty there is as to the probable supply and demand the less chance for speculation and loss in the business of distributing and marketing the crop, which is a benefit both to the producer and to the consumer.

Large manufacturing firms, agricultural implement and hardware companies, who neither buy nor sell farm products, are much interested in crop prospects. This knowledge enables them to distribute their wares economically, sending much to sections where crops are good and farmers have money with which to buy, and less to sections where crops are short and farmers will have less to spend. Few farmers realize how much is saved by an even distribution of manufactured articles according to crop prospects. If manufacturers avoid heavy losses from improper distribution, they can afford to sell on better terms, with resulting benefit to farmers.

The railroads of the country, which move crops from the farm to the market, must know in advance the probable size of the crop in order to provide a sufficient number of cars to handle it effectively and without delay. Cases are not infrequent when prices of grain at railroad stations are reduced, or there is absolutely no sale for the grain because cars are not available for shipping, the farmer thus being among the sufferers.

Prompt and reliable information regarding crop prospects is equally important and valuable in the conduct of commercial, industrial, and transportation enterprises. The earlier the information regarding the probable production of the great agricultural commodities can be published, the more safely and economically can the business of the country be managed from year to year.

Retail dealers in all lines of goods, whether in the city or in the country, order from wholesale merchants, jobbers, or manufacturers, the goods they expect to sell many weeks and frequently many months before actual purchase and shipment. Jobbers follow the same course, and manufacturers produce the goods and wares handled by merchants of every class far ahead of the time of their actual distribution and consumption. It is therefore important that they have the earliest information possible with respect to crop prospects and the probable purchasing power of the farmers.

With such information carefully and scientifically gathered and compiled, and honestly disseminated, so that it can be depended upon to be as accurate as any forecast or estimate can possibly be, and relied upon as emanating from an impartial and disinterested source, the farmers, the merchants, the manufacturers, and the transportation and distributing agencies of the country can act with a degree of prudence and intelligence not possible were the information lacking.

Scope of Crop Reports

Beginning with planting, data are gathered and reports made as to the condition and acreage of each of the principal agricultural products, such as corn, wheat, oats, rye, barley, potatoes, hay, cotton, tobacco, rice, etc. As the crops progress the prospects are reflected in monthly condition reports upon each growing crop; such reports being expressed in percentages, 100 representing a normal condition. Condition reports, expressed in percentages of a normal, when published, are coupled with a statement of the averages of similar reports at corresponding dates in preceding years (usually 10-year averages); by such comparison the condition of crops in comparison with the average condition is readily obtained. At harvest time the yields per acre are ascertained, which, being multiplied by the acreage figures already ascertained, give the production. . . .

Methods of Crop Reporting

The reports issued by the Bureau of Crop Estimates during the year include data relating to acreages, conditions, yields, supplies, qualities, and values of farm crops, numbers by classes, condition, and values of farm animals, etc. The data upon which such estimates are based are obtained through a field service consisting of a corps of paid State field agents and crop specialists and a large body of voluntary crop reporters composed of the following classes: county reporters, township reporters, individual farmers, and several lists of reporters for special inquiries.

The field service consists of trained field agents, one assigned to a single State or group of smaller States which in the aggregate corresponds in area and crop production to one of the larger States, who devote their entire time to the work and who travel throughout their territory during the crop season, personally inspecting crop areas, conferring with State and local authorities, private and commercial agencies, and others interested in crop-reporting work. Each agent supplements his own observation with reports from a corps of selected crop reporters in his territory, who report directly to him and are wholly independent of the regular crop reporters who report directly to the bureau.

In addition to the regular force of State field agents the bureau has a small force of crop specialists, one or more for each of the important special crops, such as cotton, tobacco, rice, and truck crops, possessing the same qualifications and performing the same duties as the field agents, but devoting their entire time to specializing on the particular crops to which they are assigned and traveling throughout the entire region in which they are grown. These crop specialists also have selected lists of crop correspondents reporting directly to them.

Both the State field agents and the crop specialists are in the classified service and are appointed only upon certification by the Civil Service Commission after a rigid competitive examination. They are selected for their special training and qualifications for the work and, as they acquire knowledge and experience, will become recognized authorities in crop production in each State.

There are approximately 2800 counties of agricultural importance in the United States. In each the department has a principal county reporter who maintains an organization of several assistants. These county reporters are selected with special reference to their qualifications and constitute an efficient branch of the crop-reporting service. They make the county the geographical unit of their reports, and, after obtaining data each month from their assistants and supplementing these with information obtained from their own observation and knowledge, report directly to the department at Washington.

In practically all of the townships and voting precincts of the United States in which farming operations are extensively carried on the department has "township" reporters who make their immediate neighborhood area with which they are personally familiar the geographical basis of reports, which they also send directly to the department each month. There are about 32,000 township reporters.

Finally, at the end of the growing season a large number of individual farmers and planters report on the results of their own individual farming operations during the year; valuable data are also secured from 30,000 mills and elevators.

Because of the specialized nature of the cotton crop the reports concerning it are handled separately from reports on all other crops. In addition to the regular estimates of the State agents, the cotton crop specialist, and the county and township reporters, the bureau obtain reports on acreage, yields, percentage ginned, etc., from many thousand special reporters who are intimately concerned in the crop, including practically all the ginners.

Transmission of Reports to Bureau by Correspondents

Previous to the preparation and issuance of the bureau's reports each month the correspondents of the several classes send their reports separately and independently to the department at Washington.

In order to prevent any possible access to reports which relate to speculative crops, and to render it absolutely impossible for premature information to be derived from them, all of the reports from the State field agents, as well as those from the crop specialists, are sent to the Secretary of Agriculture in specially prepared envelopes. By an arrangement with the postal authorities these envelopes are delivered to the Secretary of Agriculture in sealed mail pouches. These pouches are opened only by the Secretary or Assistant Secretary, and the reports, with seals unbroken, are immediately placed in a safe in the Secretary's office, where they remain sealed until the morning of the day on which the bureau report is issued, when they are delivered to the statistician by the Secretary or the Assistant Secretary. The combination for opening the safe in which such documents are kept is known only to the Secretary and the Assistant Secretary of Agriculture. Reports from field agents and crop specialists residing at points more than 500 miles from Washington are sent by telegraph, in cipher. The reports from the county correspondents, township correspondents, and other voluntary crop reporters are sent to the Chief of the Bureau of Crop Estimates by mail in sealed envelopes.

Preparation of Reports

The reports received by the department from the different classes of individual correspondents are tabulated and compiled and the figures for each separate State computed. After the reports from the different counties are tabulated, a true weighted figure for the State is secured by taking into consideration the relative value which the total acreage or production of each county in the State bears to the total acreage or production of the State. The weight figure showing the value of the county is applied to the acreage, yield per acre, or condition, whichever it may be, and from the totals of the weights and the extensions a weighted average for the State is ascertained. The averages for speculative crops (corn, wheat, oats, and cotton) are determined by computers who do not know the particular State to which their figures relate.

The work of making the final crop estimates each month culminates at sessions of the crop-reporting board, composed of five members, presided over by the statistician and chief of bureau as chairman, whose services are brought into requisition each crop-reporting day from among statisticians and officials of the bureau, and field agents and crop specialists who are called to Washington for the purpose.

The personnel of the board is changed each month. The meetings are held in the office of the statistician, which is kept locked during sessions, no one being allowed to enter or leave the room or the bureau, and all telephones being disconnected.

When the board has assembled, reports and telegrams regarding speculative crops from field agents and crop specialists, which have been placed unopened in a safe in the office of the Secretary of Agriculture, are delivered by the Secretary, opened, and tabulated; and the figures, by States, from the several classes of correspondents and agents relating to all crops dealt with are tabulated in convenient parallel columns; the board is thus provided with several separate estimates covering each State and each separate crop, made independently by the respective classes of correspondents and agents of the bureau, each reporting for a territory or geographical unit with which he is thoroughly familiar.

Abstracts of the weather condition reports in relation to the different crops, by States, are also prepared from the weekly bulletins of the Weather Bureau. With all these data before the board, each individual member computes independently, on a separate sheet or final computation slip, his own estimate of the acreage, condition, or yield of each crop, or of the number, condition, etc., of farm animals, for each State separately. These results are then compared and discussed by the board under the supervision of the chairman, and the final figures for each State are decided upon.

The estimates by States as finally determined by the board are weighted by acreage or other figures representing the relative importance of the crop in the respective States, the result for the United States being a true weighted average for each subject.

Method of Issuing Reports

Reports in relation to cotton, after being prepared by the crop-reporting board and personally approved by the Secretary of Agriculture, are issued on or about the first day of each month during the growing season, and reports relating to the principal farm crops and live stock about the seventh or eighth day of each month. In order that the information contained in these reports may be made available simultaneously throughout the entire United States, they are handed, at an announced hour on report days, to all applicants and to the Western Union Telegraph Co. and the Postal Telegraph Cable Co., which have branch offices in the Department of Agriculture, for transmission to the exchanges and to the press. These companies have reserved their lines at the designated time, and forward immediately the figures of most interest. A multigraph statement, containing such estimates of condition or actual production, together with the corresponding estimates of former years for comparative purposes, is prepared and mailed immediately to newspaper publications.

The crop estimates for the State and for the United States as a whole are telegraphed immediately to the Weather Bureau station director of each State, in whose office copies are printed and mailed to all the local papers in the State, so that the crop estimates of the bureau are published throughout the United States within 24 hours of their issuance.

Promptly after the issuing of the report, it, together with other statistical information of value to the farmer and the country at large, is published in the *Agricultural Outlook*,[1] a publication of the Bureau of Crop Estimates, under the authority of the Secretary of Agriculture. An edition of over 225,000 copies is distributed to the correspondents and other interested parties throughout the United States each month.

Acreage Estimates

For many years, in fact since the bureau was organized in 1862, it has been the practice to accept the estimates of acreage planted to different crops as reported by the Bureau of the Census every 10 years.[2] Then in the first year following the census the crop reporters of this bureau would estimate the acreage planted as a percentage of the

[1] Supplanted by *The Monthly Crop Reporter*, January 1, 1918.

[2] Prior to 1880 the Census did not show acreages of crops — merely production; hence in the earlier years the acreage basis was obtained by dividing the census report of total production by an estimated yield per acre.

acreage reported by the census for the preceding year; the second year following the census the acreage would be estimated as a percentage of the acreage estimated the preceding year, and so on until figures for the next census are available. Theoretically, if there is no bias or tendency to underestimate or overestimate on the part of crop reporters, the acreage estimates by this method for the tenth year after the census would agree with the acreage reported by the census for that year. A weak point in the system which has long been recognized is the fact that individual crop reports are not free from bias, and there appears to be a fairly uniform tendency to either overestimate or underestimate the acreage, the result being a cumulative error which in 10 years is apt to result in a wide discrepancy between the estimates of this bureau and the figures of the census. To illustrate, if the Bureau of the Census should report 10,000,000 acres planted to a given crop, and there should be a uniform tendency on the part of crop reporters of this bureau to underestimate the acreage of this crop an average of 2 per cent annually, this bureau might estimate the acreage as 9,800,000 acres the first year after the census, as 9,604,000 acres the second year, as 9,412,000 acres the third year, and so on until the tenth year, when the bureau's estimate for the crop would be 8,170,000. If during the 10-year period there had actually been no change in the acreage planted to the particular crop in question, and the census should again report an acreage of 10,000,000, the result would be a manifest discrepancy of 1,830,000 acres between the figures of this bureau and those of the census. Further discrepancies would appear in the yield per acre and the total yield.

At or near the close of harvest each year agents and crop reporters of the bureau estimate the yield per acre, in bushels, pounds, or tons, according to the nature of the product. The estimate of total production is readily obtained by multiplying the yield per acre thus obtained by the previously estimated total number of acres.

It will be observed that the method of estimating the yield per acre differs materially from the method of estimating the total acreage, the acreage estimate being based upon a percentage of the preceding year's acreage, thus carrying on from year to year any error made in any previous year; whereas the yield-per-acre estimate, being based upon the one year and not referring to any former year, is not affected by any error of a previous year. A constant yearly underestimate of, say, 2 per cent in the acreage will be magnified to a difference of about 10 per cent in 5 years and 20 per cent (approximately) in 10 years. A constant yearly underestimate of 2 per cent in the yield per acre will not be magnified in 5 or 10 years, but, on the other hand, in comparing one year's estimated yield with another the errors will be neutralized; that is, the effect would be the same, so far as comparative value is concerned, as though no error had occurred. In short, biased errors in acreage estimates by percentage grow from year to year; biased errors in yield-per-acre estimates neutralize each other.

The Bureau of the Census enumerates total acres and total production of crops; if yield per acre is wanted it is obtained by dividing the production by the acres. The Bureau of Crop Estimates obtains directly from its agents and correspondents estimates of acreage (as described) and yield per acre and arrives at the total production by multiplying acreage by yield per acre.

Notwithstanding the difference in methods of procedure, the estimates of yield per acre obtained by the Bureau of Crop Estimates in census years and the figures of yield per acre obtained by the census, with few exceptions, do not vary widely.

Live-stock Estimates

Practically the same difficulty is encountered by this bureau in its estimates of the numbers of different classes of live stock, *i.e.* the probable cumulative error resulting from a uniform tendency to either underestimate or overestimate and the consequent application of an erroneous percentage to the census figure the first year and to an erroneous basis in each succeeding year until the next census. A further cause of divergence between the live-stock estimates of this bureau and the figures of the census, and between any two census years, results from taking the census or making the estimates at different seasons of the year. It can readily be seen that in the case of sheep and swine the estimates cannot agree unless made as of the same date, because of the normally wide fluctuations in numbers due to natural increase during a few months in spring and the large decrease due to slaughter in the case of swine, and also from exposure and other causes in the case of sheep during the winter months.

While the Bureau of Crop Estimates has in recent years taken cognizance of the tendency to bias on the part of its field force and has endeavored to make such allowance therefor as would correct the errors involved, besides checking its estimates against the returns of tax assessors in different States and such other reliable sources of information as are available, it has felt the need for a better method of estimating acreages and live stock between the census years.

Use of Rural Mail Carriers

As an experiment, and with the coöperation of the Post Office Department, an attempt was made in the winter months of 1913–1914 to secure accurate data as to acreage planted and numbers of live stock in the State of Maryland and 15 counties in South Carolina by means of short, simple schedules left in mail boxes and collected by the rural mail carriers. In theory this plan should result in complete returns as accurate as a census, but in practice it was found that less than 40 per cent of the farmers would fill out the schedules. The experiment demonstrated that satisfactory results by this method cannot be secured without (1) a personal canvass and actual enumeration by the rural mail carriers similar to that of the census enumerators; (2) legislation making it compulsory upon farmers to supply the information requested; or (3) a long campaign through the press and other agencies to educate the farmer into the idea of furnishing information of a statistical nature regarding their business, primarily for their own benefit and incidentally for the benefit of others.

Typical Farms for Estimating Acreage and Live Stock

The experiment in utilizing the services of rural mail carriers for making an actual enumeration of acreages and of live stock having proved inadequate and unsatisfactory, even as a basis for estimating, it was decided to establish a selected list of typical farmers in each county in the United States who will agree in advance to coöperate with the department to the extent of furnishing accurate statements of acreages and live stock on their farms for a series of years.

These reports will establish a basis for comparison with the census figures and will enable the department to estimate with a high degree of accuracy the changes which take place annually between censuses. In future years it will be a simple matter to apply the rate of increase or decrease in acreages and live stock which is found to take place on the selected typical farms in each county to the total number of farms reported by the Bureau of the Census, and the results can be used to check the estimates prepared on the percentage basis under the present system. A much higher degree of accuracy will also be possible with census returns available every 5 years, as will be the case hereafter, instead of only once in 10 years as heretofore.

The "Normal" as a Basis of Condition Reports

Special consideration has been given for many years to the so-called "normal," representing a condition or yield of 100 per cent, in terms of which all the crop condition estimates of this bureau are expressed. An objection to the use of this term and what it represents, as a basis for crop reporting, arises from its apparent vagueness and the fact that the yield represented by it is different for each locality and even for each farm, thus requiring explanation in order to be understood. The principal advantage of the term "normal" is psychological in that it is based on a fundamental conception which is fairly uniform and clear in the minds of all practical farmers, from whom over 99 per cent of the crop condition reports of this bureau are received.

But little observation and experience is required to demonstrate that the average farmer thinks of his crop as "crops" and not in mathematical terms of percentages or averages, although he can readily express the estimated yield of the crop in terms of bushels, pounds, or tons. When the farmer sows the seed in spring he knows just what the field ought to yield, and if the season is favorable, he expects to harvest that yield. This expected yield is a "full crop," such as he has harvested in the past in favorable seasons. It is neither a maximum possible or even a bumper crop, which occurs only at rare intervals when conditions are exceedingly favorable, nor a medium or small crop grown under one or more adverse conditions. Neither is it an average crop, which rarely occurs because of the effect on the average of extremely low or extremely high yields in exceptional seasons. It is rather the typical crop represented by the average of a series of good crops, leaving out of consideration altogether the occasional bumper crop and the more or less frequent partial crop failure. This expected yield at planting time, the full crop that the farmer has in mind when he thinks of the yield he expects to harvest, or the typical crop represented by the average of good crops only, is the "normal," or standard adopted by this bureau for expressing condition during the growing season and yield at harvest time.

The observation is sometimes made, as a criticism of the use of the normal, that a normal crop is almost never shown in the reports of the bureau. A little reflection will show that a normal yield for an entire State or the United States is not to be expected except on rare occasions. Imagine the yields of 10 different farmers in widely scattered parts of the United States; by definition of the term normal as a "full crop," or expectation of yield at planting time, an individual will not secure a normal yield every year, or even every two years. Suppose each individual secured a normal crop on the average every three years, by the law of probability the chance of all 10 farmers getting a normal crop in the same year is 1 to 30. If returns of individuals were published, many normals would be shown, but the frequency would be less in a county average, still less in a State average, and rare in a United States average.

The crop prospect is a subject of vital interest to farmers and, like the weather, it is a perennial topic of discussion during the crop season. Almost invariably farmers speak of the prospects as fine, good, fair, or poor, and they describe the crop as "full crop," "good crop," "average crop" (meaning less than a full crop but a little better than the real average), "three-fourths of a crop," or "one-half of a crop," or less infrequently "75 per cent of a crop," "50 per cent of a crop," etc. In the South the cotton crop prospect is usually spoken of in terms of bales, as "three-fourths bale per acre," "one-half bale per acre," or "one-third bale per acre." Few farmers think of their crops in terms of exact mathematical averages or, in fact, know what the exact average really is, because very few of them keep accurate records or take the trouble to strike averages from them. It is equally true that farmers do not generally speak of crop conditions and crop prospects in terms of a normal, but when the farmer crop reporters are told that the normal is the same as their conception of a full crop, the crop which their farms ought to yield and are expected to yield in favorable seasons, and that this normal is represented by 100, they have no difficulty in clearly understanding what is meant by the normal or in expressing their estimates in percentages of normal.

Reports of crop condition expressed in percentage of normal may indicate in a general way the probable yield, but as they do not include the variations in acreage it would be impracticable to forecast total production accurately from condition estimates alone. Hence, to avoid errors in the interpretation of condition estimates by those who do not have the average figures before them, the bureau converts the condition estimates into quantitative estimates of yield per acre, which, applied to the estimated acreage of a given crop, indicate the probable total production.

The question is frequently asked why the crop estimates are not (1) based on the average crop (presumably the average for the past 5, 10, or 20 years), or (2) on the crop of the preceding year, or (3) simply estimated for the present year in terms of bushels, pounds, or tons.

The answer to the first proposition is that no "average crop" can properly be said to exist, or rather it would not correspond to any crop actually harvested, because the average for any given period is unduly influenced by the exceptionally low or high yields of abnormal seasons. In other words, the average is a fluctuating instead of a fixed standard. Furthermore, it would be exceedingly difficult to obtain satisfactory estimates of crop prospects based on average yields from farmer crop reporters, who constitute the bulk of the bureau's field force in reporting on crop conditions during the growing season. Farmers as a rule do not keep a record of average yields on their farms or for their communities. They do, of course, remember abnormally high or low yields, but they invariably leave such yields out of consideration when estimating crop prospects. If the average crop, say, for a period covering the last five years, were adopted as the standard, it would be necessary for the bureau to estimate the average condition for each month of the growing season and the average yield for each year in each county and township in the United States (over 30,000) for each of the crops included in the estimates (50 or more) and to furnish each crop reporter with the average production of each crop in his territory for use in making up his monthly estimates during the year. This would entail an enormous amount of additional work, and the average would be unsatisfactory because the smaller the unit of territory the greater would be the fluctuation in the average or standard caused by crop failures or occasional bumper yields. A single illustration will suffice to make this point clear. Taking the corn crop of Kansas as an example, the average yield of corn per acre in the State of Kansas for each of 10 years, beginning with 1903, was as follows: 20.9, 27.7, 28.9, 22.1, 22, 19.9, 19, 14.5, 23, 3.2. The average for the 10 years is 20.1 bushels; the average for the last five years is 15.9 bushels; for the preceding 5 years 24.3 bushels. On the other hand, the idea of a normal crop, or a full crop, was nearly constant, being 31.7 for the last 5 years, 31.5 for the preceding 5 years, and 31.6 for the 10 years.

The answer to the second proposition, namely, a comparison of this year's crop with the crop of the preceding year, is that while farmers remember fairly well the condition and yield of crops for the past year, they do not remember them with sufficient clearness or accuracy to be able to use them as a standard of comparison for this year. Furthermore, the crops of last year may have been abnormally high or low, and would therefore make a very poor basis of comparison. For instance, the yield of corn per acre in Kansas was 23 bushels in 1912, or 159 per cent of the yield per acre in 1911 (14.5 bushels). The yield in 1913, an abnormally dry season, was only 3.2 bushels per acre, which was 14 per cent of the yield in 1912. If the yield per acre of corn in Kansas for 1914 should be 21 bushels per acre, it would be 656 per cent of the yield of

1913. It is apparent, therefore, that the abnormally low yield of 1913 is a most unsatisfactory basis of comparison for the year 1914.

The third proposition, namely, the estimating of crops directly in terms of bushels, pounds, or tons, is sometimes advanced. The objection to this system is the difficulty that most people experience in estimating accurately, until near harvest, the number of bushels or pounds which an acre will yield, even though they may be good judges and have the field before them. Experience has demonstrated repeatedly that it is much easier to estimate proportions and differences in comparing one period with another, or the production of one year with the production of another year, or condition and prospective yield with some standard, such as a normal, than it is to estimate quantitatively what the condition or yield for a given area actually is at any given time. Any one can demonstrate this principle to his own satisfaction while looking at a shelf partly filled with books or a glass partly filled with beans. The shelf or jar becomes in each case the standard or normal represented by 100 per cent. He will probably find that he can readily estimate that the shelf or jar is three-fourths or 75 per cent full, and while he may be able to guess within 25 per cent of the actual number of books, he may overestimate the actual number of beans in the jar more than 100 per cent. So with cereals or other crops. It is relatively easy for the crop reporter to estimate the prospects as 90 per cent of the normal or other standard, but he may have difficulty in estimating within 25 per cent of the actual prospects in terms of bushels. Of course, crop estimates stated simply as percentages of a normal or other standard would not mean much, for which reason, wherever practicable, such estimates are converted into numerical statements by the bureau and their equivalents in bushels, pounds, or tons are published in comparative statements showing the figures for the previous year and the 5 or 10 year average.

This whole subject of standards or bases for crop reports has been thoroughly and repeatedly considered, both in this country and abroad. On every occasion when the subject has been considered in this bureau the normal has seemed to possess more advantages and fewer disadvantages than any other standard. The Canadian government has adopted as its basis of crop estimates the principle of the 10-year average. The 10-year average has also been adopted by the International Institute of Agriculture at Rome, and the institute is constantly urging its adoption by the adhering countries. Great Britain still uses the 10-year average as the standard, which is fluctuating. Germany and a few other European countries use the numbers 1 to 5, inclusive, to represent the condition of excellent, good, fair, poor, or very poor. In France the same gradations of conditions are symbolized by 80 to 100, 60 to 80, 40 to 60, 20 to 40, and 1 to 20. The German system results in confusion because in Germany the number 1 represents the highest condition, while in Sweden it represents the lowest condition; besides, the terms excellent, good, fair, or poor are only descriptive and are open to interpretations which interested speculators may desire to place upon them.

Accuracy of Condition Reports

The quantitative interpretation by the Department of Agriculture of condition reports of principal crops, except cotton, was begun in 1911. A review of these interpretations, or forecasts, shows that those made in June varied an average of 11.2 per cent from final yield estimates; those in July varied 9.6 per cent; in August 6.7 per cent; in September 4.3 per cent; in October 3.1 per cent. Generally forecasts made one and two months before the harvest inquiry are very close to the final estimates of yield. The above percentages do not reflect the accuracy of the work of estimating, but rather reflect the variableness of conditions affecting growing crops, which is shown by changes which take place after the dates to which the condition reports relate. The condition of a corn crop on August 1 may be normal with a forecast of 35 bushels per acre; but the crop may be practically ruined 10 days later by a devastating hot wind, and the final yield be but 2 or 3 bushels per acre. The forecasts are such figures that, based upon average conditions in past years, there is an even chance or probability that the final yield will be either above or below the figure forecast. A variation of 11.2 per cent from the June forecast does not necessarily indicate an error of 11.2 per cent in the forecast, but rather indicates an average subsequent change in condition of 11.2 per cent before harvest.

The forecasts made during the past three years, and final estimates of yield are given below:

	Forecast Made in—					Final Estimate
	June	July	August	September	October	
Corn (bushels):						
1911		25.5	22.6	23.6	23.8	23.9
1912		26.0	26.0	27.7	27.9	29.2
1913		27.8	25.0	22.0	22.2	23.1
Winter wheat (bushels):						
1911	15.3	14.6				14.8
1912	14.1	13.9				15.1
1913	15.9	15.6				16.5

	Forecast Made in —					Final Estimate
	June	July	August	September	October	
Spring wheat (bushels):						
1911	13.7	11.8	10.1	9.8		9.4
1912	13.8	14.1	15.1	15.6		17.2
1913	13.5	11.7	12.5	13.0		13.0
All wheat (bushels):						
1911	14.7	13.5	12.8	12.6		12.5
1912	14.0	14.0	15.1	15.4		15.9
1913	15.0	14.1	15.0	15.2		15.2
Oats (bushels):						
1911	27.7	23.2	23.2	23.9		24.4
1912	29.3	30.1	31.9	34.1		37.4
1913	28.8	26.9	26.8	27.8		29.2
Barley (bushels):						
1911	24.9	20.9	19.8	20.3		21.0
1912	25.2	25.6	26.7	27.6		29.7
1913	24.4	22.8	23.1	23.2		23.8
Rye (bushels):						
1911	16.1	15.5				15.6
1912	16.0	16.0				16.8
1913	16.5	16.1				16.2
Flaxseed (bushels):						
1911		8.6	7.6	7.7	8.1	7.0
1912		9.4	9.4	9.7	9.8	9.8
1913		8.7	8.3	8.4	8.7	7.8
Rice (bushels):						
1911		32.2	32.7	32.1	32.0	32.9
1912		31.7	31.9	32.7	33.4	34.7
1913		33.0	33.1	32.8	30.9	31.1
Potatoes (bushels):						
1911		81.7	71.5	74.2	79.7	80.9
1912		95.5	100.7	108.0	108.8	113.4
1913		93.1	92.0	88.1	86.7	90.4
Tobacco (pounds):						
1911		698.1	672.4	714.6	801.1	893.7
1912		844.9	820.6	817.1	816.0	785.5
1913		809.0	783.0	752.4	766.0	784.3
Hay (tons):						
1911		1.08	1.14			1.14
1912		1.40	1.49			1.47
1913		1.33	1.33			1.31
Buckwheat (bushels):						
1911			18.1	19.6	19.6	21.1
1912			19.3	21.3	21.4	22.9
1913			20.1	18.2	16.5	17.2

NUMBER OF POUNDS OF LINT COTTON (NET WEIGHT) AS ESTIMATED IN DECEMBER, ANNUALLY, BY THE DEPARTMENT OF AGRICULTURE, AND AS SUBSEQUENTLY REPORTED BY THE BUREAU OF THE CENSUS, FOR EACH OF THE SEASONS 1900–1901 TO 1913–1914, INCLUSIVE, TOGETHER WITH THE PERCENTAGE OVERESTIMATED OR UNDERESTIMATED BY THE DEPARTMENT OF AGRICULTURE EACH SEASON

CROP YEAR	POUNDS OF COTTON (000 omitted)		OVER-ESTIMATED	UNDER-ESTIMATED
	Estimated by Department of Agriculture	Finally Reported by Census Bureau	*Per Cent*	*Per Cent*
1900–1	4,856,738	4,846,471	0.2	
1901–2	4,529,954	4,550,950		0.5
1902–3	5,111,870	5,091,641	.4	
1903–4	4,889,796	4,716,591	3.7	
1904–5	6,157,064	6,426,698		4.2
1905–6	4,860,217	5,060,200		4.0
1906–7	6,001,726	6,354,110		5.5
1907–8	5,581,968	5,312,950	5.1	
1908–9	6,182,970	6,336,070		2.4
1909–10	4,826,344	4,783,220	.9	
1910–11	5,464,597	5,551,790		1.6
1911–12	7,121,713	7,506,430		5.1
1912–13	6,612,335	6,556,500	.9	
1913–14	6,542,850	6,772,350		3.4
Total 1900–1914	78,740,142	79,865,971		1.4
Years of overestimate	31,879,051	31,307,373	1.8	
Years of underestimate	46,861,091	48,558,598		3.5

The preliminary estimates of the cotton crop in December each year are checked against the monthly and annual reports of production by the Bureau of the Census. The census reports, which are presumed to be the most accurate obtainable, indicate that the Bureau of Crop Estimates has overestimated the cotton crop 6 times and underestimated the crop 8 times in the past 14 years.

The preceding tabulation gives the annual estimates of the Department of Agriculture of the production of cotton, expressed in pounds of lint, the quantity as finally reported by the Bureau of the Census, and the percentage of overestimate or underestimate by the Department of Agriculture.

As shown in the tabulation preceding, during the past 14 years the Department of Agriculture has overestimated the crop six times and underestimated it eight times. In years of overestimates the average error was 1.8 per cent; in those of underestimates the average error was 3.5 per cent; for the entire 14 years the average error was 2.8 per cent. Balancing the overestimates and underestimates shows, for the entire period, a net underestimate of only 1.4 per cent.

REVIEW

1. What is there in the first paragraph of the description of Government Crop Reports which bears upon scientific method?

2. What interest has the farmer, the manufacturer, the railroads, the salesman in crop reports? What interest have you in such reports? Write out your answer to the last part of this question.

3. What data are collected by the Government and by what method? Does this meet the demands of good statistical practices? In what special particulars?

4. How are the reports on crop estimates actually prepared? Why the great caution? Does the caution seem warranted in view of the size of the territories covered, and the number of sources of information? What bearing, if any, on the statistical side of the problem has the statement "and the final figures for each state are decided upon"?

5. What is the method of issuing the Crop Estimates?

6. What method is used by the Department to estimate acreage; to estimate yields? What effects have biased errors on both? How do the census methods differ? What is a test of wide difference in the two methods?

7. How is the estimating of live stock different from, and more or less difficult than, the estimating of acreage?

8. What application to schedule making has the experiment, conducted by the Department, to secure actual acreage by rural mail carriers; to the mandatory power of the scheduling agent; to the type of informant?

9. What relation to sampling as a statistical device has the principle of choosing typical farms for estimating acreage and live stock? to error? How large a sample is necessary? What conditions must it cover?

10. "Practical farmers" . . . furnish "over 99 per cent of the crop condition reports." Will such people understand what is meant by a "normal crop"? Why? Is this an acceptable unit? Is such a unit likely to be better understood than the unit "good crop," "full crop," "three-fourths of a crop"? Why not use the expression "average crop"? Why not compare crop condition on the basis of the previous year? Why not estimate it in terms of bushels, pounds, tons?

11. How do you interpret the figures showing the degree of accuracy of estimates to realized crop? How is this subject related, if at all, to the compensation of errors?

SAMPLING AS AN ALTERNATIVE TO A COUNT[1]

Nature of Timber Estimates

The determination of the amount of standing timber on a given area is a matter of far greater difficulty than is likely to be assumed by persons who have not been concerned with the question. To show what the difficulties are, the methods of measuring and estimating timber must be set forth in some detail.

Measurements of Lumber and Logs. — Measurements of lumber and timber in the United States are commonly made in terms of board feet. While 12 board feet make 1

[1] Adapted with permission from "The Lumber Industry Pt. I, Standing Timber," *United States Bureau of Corporations*, January 20, 1913, pp. 45–58.

cubic foot, a tree which contains 200 cubic feet of wood will make only a small fraction of 2400 board feet of lumber. A large part of the wood — all the branches and the upper part of the trunk — is not suitable for lumber, and there is always some loss in the stump. But the lumber produced is far less than 12 board feet for every cubic foot of logs suitable for sawing. The slabs, removed in squaring the log, are wholly or largely wasted; the sawdust is wasted; there is a waste because of the difficulty of sawing true; and there may be further losses on account of internal defects in the log. Each of these losses varies widely. The slabs are a larger proportion of a small than of a large log, and a much larger proportion of a crooked log than of a straight one. The waste from defects depends upon the quality of the timber, and also on the size of the pieces sawed out; for a defect which is hidden in a heavy timber or even in a 3-inch plank may come to light in a board. The waste from the difficulty of accurate sawing varies with the wood, with the character of the mill, and with the skill of the sawyer. The waste from sawdust varies with the thickness of the saw and with the size of the lumber made. Some large circular saws take out a kerf three-eighths of an inch wide, or even more. Smaller ones may take one-fourth of an inch. Many band saws and gang saws work on one-eighth or little more. A few are said to cut as little as one-sixteenth.

With a saw that takes out a quarter-inch kerf, a thickness of an inch and a quarter is required for getting out a 1-inch board; one-fifth is lost in sawdust. If 2-inch planks are sawed, the waste is only one-ninth; if timbers, say 12 inches square, the kerf is unimportant.

The contents of logs are reckoned by lumbermen in board feet. For this purpose, however, the contents are not the full volume, but the quantity of lumber that a log may be expected to make. As has been shown in the preceding paragraphs, the product depends on many things besides the length and diameter of a log. At different mills, and under different circumstances, the product of exactly similar logs may vary materially; and at the same mill, and under the same circumstances, one log may produce considerably more than another whose gross cubic contents are the same. The measurement or "scaling" of logs, therefore, is not a mathematically accurate determination of their volume, but an approximate determination of the quantity of lumber they are likely to yield.

For this purpose, lumbermen commonly use a measure called a log scale or scale stick. This is a flat stick, a quarter of an inch or more in thickness and about an inch and a quarter broad. The edges are often graduated in inches. On the faces are usually six graduations, three on one and three on the other, for six lengths of logs. These graduations run lengthwise of the stick, and show the contents in board feet, at each diameter, for logs of each length. The length of a given log is first determined, usually by the eye; the stick is then laid across the small end, and the contents in board feet are read off. The reading is supposed to give the contents of a straight, sound log; and if a log is crooked or unsound, the scaler makes a deduction according to his judgment. The measuring sticks are graduated according to tables, called log scales or log rules, which give the supposed product of logs of different diameters and lengths. Many such tables have been constructed, some from diagrams, some by mathematical formulæ, some by measurement of logs sawed and their product, and some by combinations of these methods. The *Woodman's Handbook,* published as Bulletin 36 of the Forest Service, gives

44 different rules. The differences among them are astonishingly wide. For a 16-foot log, 24 inches in diameter, the computed contents range from 268 board feet to 500; for a 12-foot log, 6 inches in diameter, from 3 board feet to 20.

For a log 12 feet long and 6 inches in diameter, most rules give values ranging from 12 feet to 20. Yet the Doyle rule, which gives only 3 feet, is more widely used than any other. It is far more inaccurate for small logs than for large, yet in great areas of the country it is used for small logs only. There is another rule of long and wide acceptance, the Scribner, which gives smaller values than the Doyle for the larger diameters and much larger values for the smaller diameters. A combination of the two has been made, by taking the smaller value for each size of logs, with very few exceptions. This combination, called the Doyle-and-Scribner rule, is the scale chiefly used in many parts of the Eastern, Southern, and Middle Western States. Mills which use the Doyle or the Doyle-and-Scribner rule, and which cut small logs, often have an "overrun" of 20, 30, sometimes, with thin saws, of 40 or 50 per cent; that is, their actual product of lumber exceeds by so much the scale of the logs they saw.

Usually the timber owned by a sawmill will give quite uniform results when handled under the same conditions. Defects are characteristic, not only of the species but also of the district where the trees grow, and by keeping records comparing the actual yield with the scale of the logs it is possible to determine the approximate relation between the two. The mill may thus compute the average overrun shown by its experience, and then reckon that its logs will in all likelihood yield approximately the same percentage above the scale.

Estimating Standing Timber. — It has perhaps been made clear enough that many uncertainties are involved in the scaling of logs. Even aside from the element of individual judgment, in allowing for defects, the mere application of the rules to straight and sound logs gives results which only approximate the product of the saw.

The estimating of standing timber introduces further difficulties. The ideal of accuracy, from the standpoint of the "cruiser" making the estimate, would be to reach the same result that would be reached, after the trees were felled, by the scaling of the logs. As just shown, this ideal falls far short of an accurate measure of the resultant lumber; but this very imperfect ideal is not approached in most estimates of standing timber. It can be approached by detailed calculation. Every merchantable tree can be counted, its diameter measured, and even its height. There may still be shrinkages between tree and log that cannot be determined beforehand. There may be concealed hollows; in some species, as cypress, there will be many. There may be much breakage in felling; this is a heavy loss in redwood. But, waiving such points, counting and measuring are enormously expensive, and such a method is hardly ever used in practice. Even if the trees are counted, the average diameter is usually estimated by the eye, and the supposed normal content of the tree of this diameter is multiplied by the number of trees. This normal content is based on the estimator's experience or on volume tables. Even the counting of trees is not only slow and expensive, but difficult. It is hard to be sure of getting them all and counting none twice.

Oftener no attempt is made to count every tree, but sample plots, perhaps of an acre each, are laid off by pacing or with a surveyor's chain, and the trees on them are counted. The result is taken as the average stand on the larger area which the samples represent.

Far the commonest method of estimating, however, is simply to look the forest over, without any counting or measuring. The examination may be made with less or greater care. The cruiser may tramp back and forth on parallel paths only a few rods apart, or he may make only one trip through a strip a mile wide. He may tramp all day without making a note, and set down at night his estimate of the area he has covered and of the whole amount of timber he has passed through.

By long experience, men learn to form judgments by these rough methods, which, on an average, approximate fairly the scale of the logs. The general tendency is to estimate below the truth, because the estimator desires to be "safe"; that is, not to have his estimate subsequently proved too large by other cruisers or by the results at the mill. To overestimate reflects on the cruiser. The owner will not complain if the cut shows more timber than the estimate, but he will be displeased — especially if he bought on the estimate — if the cut shows less.

Moreover, an estimate which is accurate according to the customs of one time will be inaccurate according to those of another, because the standards of merchantable timber change. With higher prices for lumber, more logs are brought to the mill from the same tract and more board feet of lumber are made out of the same log, because the manufacturer is able to sell some low-grade lumber not previously marketable. Again, some species formerly regarded as worthless and not included in estimates become valuable with higher prices and increase the estimates of merchantable timber by their amount. This has been true of every timber region in the past, and as values rise and timber is cut closer in the future, estimates will rise far above those which are used to-day.

If two estimates of the same tract, made at the same time, do not differ more than 10 per cent, they agree quite as closely as can be expected. Good estimators often differ 25 per cent, and sometimes even 50 per cent. An important tract of pine in northern Minnesota was examined by three companies in 1909, with a view to purchase. One estimated it at 125,000,000 feet, and another at 135,000,000. The seller's estimate was 170,000,000, and on this basis the third company bought it. The purchase was made, however, against the opposition of a member of the buying company, who is reputed to be one of the best timbermen in Minnesota, and who estimated the tract at from 95,000,000 to 110,000,000. The accepted estimate exceeded his by more than 50 per cent, and if the mean of his figures be taken as representing his opinion the independent estimates of other prospective buyers exceeded his by 20 or 30 per cent.

The following table shows the average results, by years, of two series of estimates — first, those made by a company in the North Carolina pine region for purposes of purchase; second, those made by the State of Minnesota on timber owned by the State for purposes of sale. The quantities given as cut represent the scale of the logs; the quantity of lumber actually sawed was materially greater.

The southern company usually paid a lump sum for a tract, and the prices it offered were fixed on the basis of its estimates. It would try to get a fair idea of the timber it was buying, but would wish to err rather on the conservative than on the liberal side. The State of Minnesota did not sell its timber at so much for a tract, but at so much a thousand, and the payments were determined by the quantity of logs scaled.

Estimated Amounts of Timber on Certain Tracts, Classified by Year of Purchase, with the Amounts Cut Therefrom

Year	Timber Bought by a Southern Company			Timber Sold by the State of Minnesota		
	Estimated	Cut	Cut, Per Cent of Estimate	Estimated	Cut	Cut, Per Cent of Estimate
	M feet	*M feet*		*M feet*	*M feet*	
1886				24,540	37,859	154.3
1887				40,472	34,021	84.1
1888				20,400	27,488	134.7
1889				33,040	49,952	151.2
1890				52,130	63,681	122.2
1891				78,710	176,784	224.6
1892				29,135	72,680	249.5
1893				23,795	36,791	154.6
1894				33,870	42,856	126.5
1895	2,550	3,774	148.0	27,403	41,010	149.7
1896	460	1,249	271.5	2,600	1,758	67.6
1897	41,075	53,508	130.3	51,322	68,598	133.7
1898	25,648	31,718	123.7	30,643	42,688	139.3
1899	29,355	37,198	126.7	4,035	3,484	86.3
1900	4,575	5,997	131.1	69,128	71,958	104.1
1901	4,485	5,539	123.5	25,400	29,565	116.4
1902	4,550	5,487	120.6	52,710	53,922	102.3
1903	3,505	3,525	100.6	70,875	82,045	115.8
1904	24,936	23,534	94.4	32,900	36,718	111.6
1905	10,142	9,556	94.2	68,078	105,970	155.7
1906	10,085	9,917	98.3	26,705	47,227	176.8
1907	590	824	139.7	22,795	27,105	118.9
1908	1,200	1,353	112.8	(1)		
1909	2,985	3,426	114.8	2,165	5,541	255.9
Total	166,141	196,605	118.3	822,851	1,159,701	140.9

Some of the earlier purchases of the southern company stood several years between buying and cutting, and if the timber was immature the quantity may have increased somewhat by growth. This element is believed to have been of minor importance, however, and it does not enter

[1] No Sales.

in the case of the Minnesota timber. That was usually cut within two or three years after the estimate was made; and in any case the timber was mature, and the decay of the old trees probably balanced the growth of the young.

Under these circumstances, the scale of the logs from the Minnesota timber, taking all the sales of each year together, was usually from 10 to 60 per cent above the estimate, with an average of 40 for the whole. The sales of 1896 cut only two-thirds of the estimate; those of 1892 and 1909 cut $2\frac{1}{2}$ times the estimate.

In the case of the southern company, reckoning its purchases by annual aggregates, the purchases of most years, so far as they have been cut, have produced logs exceeding the estimates by from 10 to 40 per cent, with an average of 18 for the whole. Three years show a shortage of from 2 to 6 per cent, and one rather small lot went above $2\frac{1}{2}$ times the estimate.

The variation is greater on particular tracts than on yearly aggregates. The following table shows the estimates and the scale of the logs, in detail, for the several tracts bought by the southern company in 1909:

Estimated Amounts of Timber on Certain Tracts Bought by a Southern Company in 1909, and the Amounts Cut Therefrom

Estimated	Cut	Cut, Per Cent of Estimate	Estimated	Cut	Cut, Per Cent of Estimate
M feet	*M feet*		*M feet*	*M feet*	
75	80	106.7	675	443	65.6
40	35	87.5	200	161	80.5
20	15	75.0	425	500	117.6
550	1,059	192.5			
1,000	1,133	113.3	2,985	3,426	114.8

On the whole year's purchases the scale of the logs varied only 15 per cent from the estimates; but on particular tracts the result ranged from 34 per cent below the estimate to 92 per cent above.

The following table, except the percentages, is taken from the report of the Commissioner of the General Land Office for 1910, page 15. It gives the results of logging on ceded Chippewa lands in Minnesota, grouping the tracts according to date of sale. Payment is based on the amount actually cut.

ESTIMATED AMOUNTS OF TIMBER ON CERTAIN CEDED CHIPPEWA LANDS IN MINNESOTA, GROUPED BY DATE OF TIMBER SALE, WITH THE AMOUNTS CUT THEREFROM

DATE OF SALE	GOVERNMENT ESTIMATE	CUT	CUT, PER CENT OF ESTIMATE
	M feet	*M feet*	
March 2, 1903	13,636	26,816	196.7
December 5, 1903	223,921	308,637	137.8
December 28, 1903 . . .	169,308	296,155	174.9
November 15, 1904 . . .	146,560	168,113	114.7
November 17, 1904 . . .	9,718	18,786	193.3
July 17,1907	2,056	3,754	182.6
March 15, 1910	2,169	2,189	100.9
Total	567,368	824,450	145.3

On the whole quantity the log scale exceeded the estimate by 45 per cent. On the tracts sold November 17, 1904, and on those sold March 2, 1903, the log scale was nearly double the estimate.

Professional cruisers keep as well informed as possible on the relation between their estimates and the results shown in cutting the timber, and thus modify their judgment with experience. This is especially true in the first years of their work as cruisers, or when going from one timber region to another of very different character, or during periods of marked change in the standards of merchantable timber. On first going from the Lake States to the Pacific coast, cruisers made estimates far below the truth, because the stands per acre were so enormous that men accustomed to eastern stands could not grasp or accept them. It is only during recent years that estimates for western timber have been made close to the actual yield.

Methods Followed in the Investigation

In the preceding section, the effort has been made to show how far from exactness is the art of estimating timber. Even when the estimates are made with what is considered reasonable care, for the purpose of purchase or sale, they are uncertain. In naming offhand the probable contents of a tract which he has never carefully examined but has only a general knowledge of, a man will of course do worse, on an average, than in giving an estimate on a tract which he has just examined for the purpose. The most experienced lumberman can know but a comparatively small area by careful examination. When he undertakes to make a general estimate for a district, even for a few townships, he must usually depend partly on general observation and partly on the opinions of others.

Most individuals and corporations owning important tracts have had fairly good estimates of their timber made, either recently or in earlier years, and in the latter case they usually have a fairly definite opinion, based on the results of cutting or on general information, regarding the per cent by which the old estimate should be increased to make it approach present-day standards of merchantable timber. The owners of timber, cruisers, loggers, timber dealers, and the responsible employees of timber and lumber companies are often well acquainted with the approximate amount of timber in holdings other than those in which they are directly interested, and also well informed regarding the probable total amount of timber in certain survey townships or other subdivisions of a county, or in the county as a whole. Thus, there exists in the records of timber owners and in the minds of men a basis for arriving at the approximate amount of timber in a State and in the country. The accuracy of the results which may be obtained from these sources depends largely on the willingness and truthfulness with which the informants give the information they possess, and on the perfection of the methods by which this information is gathered in detail by small areas and is studied.

The only better method would be a careful examination of the timbered area by public officers. The result would still be a collection of opinions, not of mathematical determinations; but the opinions would have more value, other things being equal, in proportion as they were based on more careful and detailed examination of the timber. By the expenditure of time and money, they might be raised to any degree of accuracy, up to the point where they should represent a count and measurement of every tree.

A count and measurement of all merchantable trees, however, or even a count without measurement, would, of course, not be thought of. Such work is so expensive that most timber is bought and sold without it; and a procedure which men cannot afford to use for their guidance in buying and selling is far too expensive for any statistical inquiry. The only proposal which could be thought of would be an estimate by general observation, perhaps supplemented, where the forest was practically unbroken, with a count on sample tracts. The cost of such an estimate would vary with the minuteness of it, but the roughest canvass that would be worth making would be a matter of some millions of money and some years of time. Even if money were unlimited, it is not likely the work could be tolerably well done in ten years for lack of men. The estimating of timber is an art acquired by much practice. The men skilled in it are few, and they are employed in current business. They could hardly be diverted in the necessary numbers to an official investigation. Furthermore, such a plan would give information on the total amount of timber only and nothing regarding the ownership of it. To provide such data, it would be necessary to first obtain records of the ownership, and then to make the observations separately for each holding, which would greatly increase the expense and the time.

Methods Adopted. — The problem before the Bureau of Corporations was to provide a plan which would give, within reasonable expense and time, as accurate information as the nature of the problem allowed regarding all large holdings separately, and regarding the scattered small holdings as a whole, in order to determine the proportion between the timber owned in holdings of certain specified sizes and the total timber in the country. Under the plan adopted, the investigation of the amount of timber in all small holdings proceeded side by side with the investigation of the essential facts regarding large holdings, in such a way that the latter checked and contributed to the former.

The work was guided by the following principles:

1. The available resources would not permit the employment of estimators with a view to the examination of timber. Any estimate must therefore be based on data already existing, in records or in the minds of men.

2. The estimate of the timber on each area should be derived from the records or the opinions of those most familiar with it, and as many records and opinions as possible should be obtained regarding it, in order to give a constant check on the work and to enable the Bureau to arrive at the best estimate from a thorough study of all the available evidence in detail.

3. A separate report should be made for each holding of 60 million board feet or more. Information regarding each such holding should be obtained from as many sources as possible.

4. For the total timber in holdings of less than 60 million board feet, the best local evidence must be relied on. Estimates should be obtained for the smallest possible units of area, and the opinions of each authority should have special weight for the neighborhood which he knows best. . . .

A few reports were obtained by mail, but for nearly all owners the schedule was filled by special agents of the Bureau visiting the informants. With regard to the amount of timber, the essential items are these: The number of acres, the exact location of the land, and detailed estimates . . . which would enable the Bureau to judge the accuracy of the estimate. All States of the investigation area except Virginia, North Carolina, South Carolina, Georgia, and Texas are surveyed under the rectangular-survey system, and there it was possible to show the exact location of the timber holdings. . . . In Virginia, North Carolina, South Carolina, Georgia, and Texas, maps or blueprints showing the exact location of the land were obtained wherever available, and other holdings were located descriptively as accurately as possible by political subdivisions of the county and by the relations of the holdings to towns, railroads, streams, etc. The largest holders in these States usually have maps showing the exact location of their lands. In the rectangular-survey States, the agents did not secure the exact location in every case, and a relatively few holdings were located only by counties or as in certain survey townships; but in nearly all cases the exact location was obtained. . . .

Field work was begun by sending agents into the lumber centers, which are the headquarters of many of the largest owners of standing timber. The reports from them were tabulated by counties, showing for each holder in the county the number of acres, amount of timber, and stand per acre, and the land was platted on county maps with a different symbol for each holding. In the five States without the rectangular survey, the location of holdings could be shown exactly wherever blueprints or maps had been furnished, and in other cases only descriptively. With these records of information already obtained, an agent of the Bureau was sent into every one of about 900 counties in the investigation area. His instructions were to seek out every reliable local informant and secure all available information that would verify or correct the reports already obtained, to secure a separate report on each remaining holder in the county who had as much as 60 million feet in the United States, and to secure data in as much detail and from as many different sources as possible regarding the total timbered acreage and the amount of timber in all holdings not separately reported, including the small scattered tracts sometimes referred to as "farmers' woodlots."

By adding the holders separately reported; as he obtained them, to the county map above mentioned, the agent was able to proceed systematically in obtaining data regarding all land within the timber line of the county. For many of the counties in the Southern Pine Region and in the Lake States it was not practicable to obtain these estimates on the timber in the county or subdivisions of it, such as survey or political townships, exclusive of the reported holdings of at least 60 million feet. In such counties it therefore became necessary to secure the estimates on the total timber in the county or a subdivision of it, and then to obtain the amount in holdings of less than 60 million by subtracting the total timber reported in holdings of that amount or more. This was especially true in the five States not having the rectangular survey. But in the five States of the Pacific-Northwest, containing the great supply of timber, the estimates of the total in holdings not separately reported were obtained almost without exception by the use of maps showing the location of reported holdings. The informants, with these maps before them, made general estimates on the timberland not so platted.

All holdings of less than 60 million feet for which separate information was easily available, or which were made the subject of inquiry through belief that they might be above the limit, were separately reported, and were then tabulated and platted like the larger holdings. The proportion of the total timber in holdings of less than 60 million thus separately reported is very high in some States, and this increases the accuracy of the work.

For the timber in holdings of at least 60 million feet, the primary reliance was on the estimates of the owners or their representatives. But these estimates were not treated as necessarily conclusive. Many of them were made years ago, and omitted kinds or sizes of trees that were not then accounted merchantable, but are so accounted now. Many were admittedly only rough approximations. A very few owners, especially such as have borrowed money on their timber, were disposed to claim more than they possessed; very many holders did not wish the Bureau to know how large their holdings were. Some purposely gave erroneous information; others avoided the issue by giving access to old records which did not show the amount of timber under present standards, and withholding more recent records and facts within their personal knowledge. Agents were instructed to watch for errors from all these causes and to gather such evidence as might be available for correcting them. The owner's estimate was taken as prima facie evidence of the amount of his holding, but it was checked, wherever possible, with the estimates of other competent persons, such as former owners, timber estimators who had examined the tract, business associates, and others.

While the platting of the land owned by each holder and the replatting of it on county maps required a great deal of time, the work was absolutely essential to the investigation. An informant who would have understated the acreage owned was deterred therefrom by having to show its location. Through the use of the plats, other men could be interviewed regarding the amount of timber on the holding or such subdivisions of it as they were familiar with. Occasionally land not reported by the owner but otherwise indicated as owned by him could be added through further inquiry and the owner's supplemental statement. Again, the use of plats prevented duplication, and made it possible to say positively that a holding reported under one name was or was not the same, wholly or in part, as a holding reported by another agent under another name. Such duplication results from transfer of ownership during the inquiry and from occasional uncertainty on the part of local informants regarding the owner of record. An estimate on a given tract might be obtained from the corporation or individual who owned it at the time, and some months later, in another State, the estimate might be obtained from a corporation or individual who had bought the tract in the meantime.

As has been said, the field work was begun in the lumber centers, where men may be found who own timber from Florida to Washington. The agents were invariably instructed to report information from every authoritative source, on all timber wherever situated. On the holding of an Oregon corporation, for example, one estimate might be obtained from the manager at the mill, another from the treasurer at Portland, and another from the president in Wisconsin. Sometimes such estimates differed widely. There might be additional estimates from persons holding less responsible positions in the company, or wholly unconnected with it, such as cruisers who had examined the timber for the present owner or for others. In some States, notably Washington, estimates had been made by public officers for purposes of taxation, and these records were carefully considered. All the available estimates for each holding separately reported were transferred in the office from the original reports to a single tabulation sheet, so that they could be readily compared. Then the evidence was carefully weighted, with due regard to the position, means of knowledge, and apparent credibility of each informant. The estimate finally set down was a result of the consideration and balancing of testimony from many sources, often conflicting. In every case, an effort was made to arrive at the best possible judgment; but the care and time devoted to the effort were increased with the importance of the specific case.

Before determining the final estimates placed on these "company sheets" (each company sheet showing the estimates for that particular holding, by counties) preliminary tables had been prepared for each county, giving the number of acres, estimate of timber by species, and average stand per acre, for each separately reported holding in that county. These preliminary "county tables" threw much light on the estimates for particular holdings, for with the help of the county maps the average stands reported by neighboring owners could be compared with a view to detecting abnormal variations. Again, the county tables of particular holdings were a valuable aid as a check on the general estimates for the unenumerated holdings and on the total timber in the county. Over large areas, the average stand given for the holdings of less than 60 million feet was compared, township by township, with the stands reported by the separate holders above that limit.

When the data gathered by the field work had been collated in the office, agents were sent out a second time over practically all the timber area in the five States of the Pacific-Northwest, to verify and correct the results. The agents now had in their hands a digest of all reports previously made, and the conclusions reached in the office, together with a statement of the principal points on which there was uncertainty. The maps on which the separate holdings had been platted showed how the holdings were locally related to each other; which lay side by side and which were intermingled. Sometimes the map and the tables showed that an owner's land was closely associated in location with that of others who had reported two or three times as much timber per acre. When this appeared, the agent sought for the explanation. In some cases he was satisfied that all the estimates were honestly made and reasonably accurate; in others he obtained admissions from the owners themselves, or good evidence from other sources, that some of the estimates first given were far from the truth. This second visit to the Pacific-Northwest was necessary in greater part because of the unwillingness with which many of the most important owners there had met the Bureau's request, some of them giving data which were admitted on the second visit to be incorrect; and in lesser part because of the very marked change in the standards of merchantable timber in that region. This has largely destroyed the value of the estimates made several years ago, and many of the estimates first given to the Bureau were of this kind. The aim was to get sufficient evidence to correct all estimates to an approximate agreement with present-day standard of merchantable timber.

This second period of field work in the five States of the Pacific-Northwest not only overcame these two difficulties, for the most part, but also increased the general accuracy of the work so that the data for that region are believed to be more reliable, according to current standards, than those for either the Southern Pine Region or the Lake States. In the course of the investigation, the Southern Pine Region was taken up first, then the Lake States, then the Pacific-Northwest, and after that the second visit to the last region. The methods used developed toward perfection as the work went on, and the agents became more and more experienced, and this played a very important part in overcoming the greater difficulties in the West.

REVIEW

1. The accuracy of the estimate of lumber from logs seems to be conditioned by the measuring scale, the diversity of conditions, and the personal equation. In what respects is each of these involved? Do the "errors" due to these tend to compensate each other?

2. What are the methods of estimating standing timber? How feasible is a count of trees and scaling of the logs? Is there an element of bias in any of these methods? Why or why not? Can estimates be scientific? Why?

3. State the principles which guided the Bureau of Corporations in making an estimate of standing timber. What methods were followed? Might these be called "drag-net" methods? Why? Does the method of "balanced testimony" seem to you good? Good for other purposes? What? Illustrate.

4. What principles of statistical methods does this extract illustrate? Would these be true of other problems of sampling and estimating?

5. Just how important in your judgment is the *personal* element in this problem?

6. What standards of accuracy seemed to be aimed at here? Is accuracy always a relative term? Why?

SAMPLING IN THE DEVELOPMENT OF MARKETS [1]

The business man must first realize the intricacy of the problems he has to solve. He must analyze his market. . . . The business man faces a body of possible purchasers, widely distributed geographically, and showing wide extremes of purchasing power and felt needs. The effective demand of the individual consumer depends not alone upon his purchasing power but also upon his needs, conscious or latent, resulting from his education, character, habits, and economic and social environment. The market,

[1] Adapted with permission from A. W. Shaw, *Some Problems in Market Distribution*, Harvard University Press, 1915, pp. 100–119.

therefore, splits up into economic and social strata, as well as into geographic sections.

The producer cannot disregard the geographic distribution of the consuming public. He may be able to sell profitably by salesmen where the population is dense, while such method of sale would be unprofitable in a region where there is a sparse population. If he bases a judgment upon the average cost of selling by salesmen for the whole market, he may easily go wrong, since the average might show that the use of such an agency was on the whole profitable, although in some sections entering into the calculations the use of salesmen was actually unprofitable. Again, it might be economical for the distributor to establish his own branch stores in the denser urban centers, while in the sparsely populated regions he could most profitably distribute his product through the regular channels.

If, then, a sound system of distribution is to be established, the business man must realize that each distinct geographic section is a separate problem. The whole market breaks up into differing regions.

Equally important is a realization of what may be termed the market contour. The market, for the purposes of the distributor, is not a level plain. It is composed of the differing economic and social strata. Seldom does the ordinary business man appreciate the market contour in reference to his product. Yet obviously the success of the producers of trade-marked hats depends upon a realization of this element of market contour. The distributor of a staple hat at $3.00 appeals to different economic and social strata, faces different considerations, and finds different selling methods necessary, as compared with distributors selling a $5.00 trade-marked hat, or those distributors selling $4.00 or $6.00 trade-marked hats. Differences in economic and social strata to be reached are as important as differences in geographic location and density, if a sound system of distribution is to be worked out.

Take the distributor who seeks to map out a selling campaign for a Catholic publication. It is essential that he take into account not merely the geographic distribution of the Catholic population in the United States, the regions where it is relatively dense, and the regions where it constitutes a small element in the population, but also he must take into account the distribution of that population through the economic strata of society. A method of distribution successful in New Orleans, where the Catholic population is dense and spread through all economic strata of society, might well fail if applied in Maine, where the Catholic population is relatively sparse and found mostly in the lower economic strata. A careful analysis of his market, then, by areas and by strata, is the first task of the modern distributor.

Choice of Agencies in Distribution

Nor does the merchant-producer ordinarily realize how intricate is his problem as to the agency or combination of agencies that will be most efficient in reaching his market. . . . The business man often adopts one method and becomes an advocate of it, disregarding entirely other methods. While the method adopted may be more efficient than any other *single* method, it is apparent that a method which is relatively efficient in reaching one area may be inferior to another method in reaching another area. And so a system of distribution which has proved very effective in reaching one economic stratum may be relatively inefficient when employed to reach a different economic stratum in society.

The problem, then, of working out the most effective combination of agencies is a most complicated one. Each distinct area and economic stratum must be treated as a separate problem, and, moreover, the economic generalizations embodied in the law of diminishing returns must be taken into account in choosing that combination of selling agencies which will give, in the aggregate, the most efficient organization of the market.

Thus the distributor may find as he extends his operations in his immediate territory, geographically, that his selling cost steadily decreases, but that when he further extends his market the selling cost increases. He may find that in more distant areas selling by salesmen ceases to be profitable, and there he will perhaps establish a more economical system of selling by a combination of salesmen and circular letters. That is, he may reduce the number of visits by salesmen by one half, and supplement their efforts by a series of circular letters or more personal correspondence. In even more distant areas, it may be necessary to eliminate the salesmen entirely and to sell only by direct advertising. . . .

A sound selling policy, then, must be built up on a careful analysis of the market by areas and strata, and upon a detailed study of the proper agency or combination of agencies to reach each area and stratum, taking into account always the economic generalizations expressed in the law of diminishing returns. It must also take into account not only the direct results obtained from the use of one or the other agency over a short period, but also the less measurable results represented by the unexpressed conscious demand and subconscious demand, which go to aid future selling campaigns.

All this tends rather to give a general sense of direction

than to serve as a practical and tangible method of handling a specific problem of distribution. A clear grasp of the problem through a careful analysis is the first step in solving difficulties. To suggest any cure-all or even any panacea for the existing maladjustments in distribution, even were it possible, is not the purpose of this paper. The very complications revealed by analysis indicate the inadequacy of any single remedy. But it is possible to face the problem of remedy as well as of diagnosis in a scientific spirit, — to introduce what may be termed the "laboratory method."

Laboratory Study of Distribution

The crux of the distribution problem is the proper exercise of the selling function. The business man must convey to possible purchasers through one agency or another such ideas about the product as will create a maximum demand for it. This is the fundamental aim, whatever the agency employed. Hence this is the point where a scientific study of distribution must first be applied. How is the business man to determine what ideas are to be conveyed to the possible purchaser and what form of expression is best adapted to such conveyance?

Here, as elsewhere in distribution, the ordinary business man is to-day working by rule of thumb. He *guesses* at the suitable ideas and forms of expression, and gambles on his guess. On the basis of his *a priori* selection of ideas fitted to build up a demand for his product and of a form of expression suited to convey the ideas effectively, he invests tens, even hundreds of thousands of dollars in a selling campaign.

The more able business men, to be sure, seek to determine those facts about their goods that will attract the attention of the possible purchaser and awaken in him the desired reaction that is, a desire for the article. They study in a general way the points of superiority in quality and service possessed by their products as compared with other goods of like kind.

They also seek guides as to the form in which the ideas should be conveyed, in the general principles of style, all based on the fundamental notion of conserving the prospective purchaser's mental energy by cutting down the friction of communication. They know, for instance, that they should use short familiar words expressing their exact shade of meaning; that they should give preference to figurative language; that they should suggest a concrete image only after the materials of which it is to be made are conveyed; that they should avoid abstractions and generalizations where possible; that when they are suggesting the reaction desired their language should become quick, sharp, and compelling.

These things the more efficient business men know and apply. But all this is *a priori*. The need is for a method of practical test that will enable us to try out selling ideas and forms of expression, under laboratory conditions, as it were, before the investment of thousands and hundreds of thousands of dollars is staked on the success of the selling campaign.

Mention has been made of the annual expenditure of not less than a billion dollars in advertising. Unquestionably an extremely large percentage of this is wasted. This means not merely individual loss, but social loss. It is a diversion of capital and productive energy into unprofitable channels.

The causes of this waste are numerous. The commodity in question may be one not possessing those elements of quality and service which constitute the basis for a demand on the part of the consuming public. If the goods advertised are not adapted to satisfy a need, conscious or subconscious, of consumers, the advertising cannot be effective. Attempting to sell a thing that nobody needs is wasted effort.

Again, the medium used for the communication of the ideas about the goods may not be one that reaches the particular economic or social stratum in which possible purchasers of the commodity lie. Hence the ideas fail to create a demand because they do not reach those in whom a latent need for the commodity exists.

Another important cause of advertising waste lies in the failure to take advantage of aroused demand. The distributor often fails to give proper attention to the matter of the physical supply of his product. There results a considerable leakage in demand from the inability of persons in whom a demand has been created to obtain the goods at the time when desired.

But the great cause of waste is probably the fact that the ideas about the goods, or the form in which those ideas are conveyed to possible purchasers, prove ill-adapted to secure the desired reaction, and thus to create in the consumer an effective demand.

If we can apply to this pressing problem of advertising waste methods of study which have proven efficient in other fields, the gain is clear. The engineer does not choose material for a bridge by building a bridge of material and waiting to see whether it stands. He first tests the material in the laboratory. That is what the business man must do.

The statistician turns in his problems to the law of averages. He is familiar with what are termed mass phenomena. He knows that he can learn something of the average height of a body of people by studying the heights in a group of a few thousands of people drawn at random from the larger body. Provided that the smaller group is so selected as to insure that it is typical of the larger body, and provided the group is large enough to render the law of averages applicable, the statistician knows when he has determined the average height of the smaller group that it will roughly coincide with the average height of the larger group.

This method of study can be applied by the business man in testing the ideas and forms of expression to be used in a selling campaign. In direct advertising, the mailing of selling letters, circulars, or catalogues to prospective purchasers to draw from them an order for goods as an evidence of awakened demand, you have a stimulus and response adapted to direct statistical measurement. The number of responses per thousand communications can be determined. Here is the agency that the business man can employ in testing, under what are equivalent to laboratory conditions, the ideas and forms of expression that seem to him best adapted to awaken a demand for his product.

Suppose the manufacturer of a food product is planning a campaign to reach, not the consumer, but the grocers of the country. Now the whole body of dealers, large and small, handling groceries numbers something like 250,000. Let the distributor, after working out a set of ideas and forms of expression which seem to him likely to be effective in arousing the desired demand, test this material by mailing it to say 1000 grocers. The group selected must be large enough to give typical results and it must be so selected as to be representative in character of the whole body of grocers.

Granting these elements, the distributor can determine the number of responses from the 1000 grocers to whom the communication was sent, and can estimate from that result the average response per thousand of communications that would have been obtained if the same ideas in the same form of expression had been conveyed to the whole body of 250,000 dealers in groceries in the country. He can then test by means of direct mailing to another group of 1000, a varying set of ideas or varying form of expression. And so on with other modifications of the selling material. Thus it will be possible to determine what ideas, in what arrangement and in what form of expression, are most effective to arouse the desired demand.

That the plan suggested is practical is indicated by the results of such an intensive study presented in the table below. Here are shown the results of "tests" and the results of complete mailings. The tests here covered only one stratum of society, a mailing list of bankers being used. The purpose of the selling material mailed was to obtain orders for certain publications. Various forms of "copy" were tested by mailing, usually to 500 names on the list. Where the return on any test exceeded the minimum standard of twenty orders per thousand communications the material was mailed to the complete list. In only one case did the complete mailing fail to show an average return per thousand communications substantially the same as that derived from the test mailing. In the case of Test D^1, mailed September 15, 1909, the return is clearly out of proportion to the results from the mailing. The same material mailed on the same date, however (Test D^2), gives for a similar small group a return much closer to the results obtained from the final mailing. When a minimum standard as low as twenty is used, and the test group numbers only 500, there is danger that the average will be disturbed as by one individual sending in several orders. The larger

BANKERS' TESTS

Minimum Standard = 20 per M

Material Mailed	Tests				Mailings			
	Date	No. of Pieces Mailed	Total Orders Received	No. per M	Date	No. of Pieces Mailed	Total Orders Received	No. per M
	1909				**1909**			
A^1	3/30	500	3	6				
A^2	3/30	500	5	10				
B^1	8/13	500	6	12				
B^2	9/13	500	3	6				
C^1	9/15	500	4	8				
C^2	9/15	500	3	6				
D^1	9/15	453	6	25	9/27	19,943	360	18
D^2	9/15	500	18					
E	9/16	500	7	14				
F^1	9/21	500	24	36	11/23	16,511	589	35
F^2	9/21	500	12					
G	10/18	1,000	30	30	11/28	21,790	643	29.5
					1910			
H	11/16	500	11	22	1/24	6,554	165	24
					1/24	16,039	390	
	1910							
I	4/11	500	12	24	5/5	6,810	145	25
	4/11	500	12		5/4	12,154	336	

NOTE. — Where the same letter appears with different exponents under "material mailed" it indicates that on the test mailing results were kept separately for the same material mailed to two small groups.

the test group the more exact an index will it give as to the results which will be obtained from a complete mailing.

This method of studying ideas and forms of expression in direct advertising would be important, even though its usefulness did not extend beyond direct advertising. It would permit one to guide a widely extended direct advertising campaign by an investigation relatively inexpensive.

But the importance of the method described does not end with direct advertising. Remember that the root idea is the same, whatever the agency for selling employed. Selling is accomplished by communicating to the possible purchaser ideas about the goods calculated to stimulate in him a desire for the goods. These ideas may be communicated through middlemen, salesmen, general advertising, or direct advertising. Since the ideas are the same, whatever the agency for communication, the business man can determine in his direct selling laboratory, what ideas and in what combination are the most effective selling material. He can then carry over into his selling by other agencies the knowledge there obtained.

Suppose an extensive campaign through periodicals is under consideration. The distributor contemplates spending perhaps hundreds of thousands of dollars upon advertising in certain periodicals. What can the "distribution laboratory" do to determine the ideas to be conveyed and the forms of expression to be used to create the desired demand? Now the circulation of a periodical to be used may run into the hundreds of thousands or even into the millions. The business man wishes to test the response that will result from the communication to this enormous body of subscribers of certain ideas expressed in certain forms. Not only can he work out the most effective ideas, the most effective arrangement, and the most effective forms of expression through the agency of direct mailing, but he can even test the final "copy" itself, just as it will appear in the periodical, by mailing it directly to relatively small groups.

Moreover, he can test the response to it found in differing strata of society. Ideas adapted to build up a demand for a commodity in one economic or social stratum may prove ineffective when dealing with another. The importance of this method lies in the fact that most periodicals circulate within certain fairly well-defined economic and social strata. The ideas and forms of expression that are most effective in one periodical hence may be relatively ineffective if used in another that reaches a different stratum.

Equally important is the application of the suggested method of study to selling through salesmen. The more progressive business men to-day train the salesmen in a certain basic "selling talk." That is, certain ideas, arranged in a certain order and expressed in certain forms, are impressed upon them as likely to build up a demand for the article on the part of possible purchasers. The basic "selling talk" is not, of course, repeated parrot-like by the salesman, but it does serve as a foundation for his talks to possible buyers.

Here again the laboratory idea can be applied. The whole structure of the selling talk can be built up on the ideas, order of arrangement, and forms of expression established as the most efficient in creating demand through the medium of direct advertising. One need but appreciate the fundamental identity of the selling function, through whatever agency exercised, to realize that the results obtained in experiments in direct advertising can be carried over to selling by salesmen.

Note, too, that the general principles upon which the "testing" method depends, apply when we seek to study the possibilities of the whole market by the intensive cultivation of one section of it. A localized selling campaign, narrow in extent, will give relatively exact data from which the possibilities of a nation-wide campaign of like character may be judged. Obviously, if our law of averages holds good, we may carry over the results obtained in one section to other sections, and hence at small cost guide a widespread campaign.

The exact data that can be obtained through such "testing" methods permit a more scientific consideration of the decreasing returns obtained if one agency is used beyond a certain point. Hence a better combination of agencies is possible, with a view to the greatest aggregate efficiency.

The Effect of Different Price Policies

When a business man contemplates putting a new product on the market, a serious problem is the price at which it shall be sold. In the introduction of a safety razor, for instance, at what price is it to be sold? In such a case the business man seeks to determine which price will give him the best net return, all things considered. Now the method of study developed above will permit the business man to determine by actual test the effective demand that can be built up at different price levels in different economic and social strata. Hence he can fix the price on the basis of relatively exact data, rather than on a mere guess.

Again the laboratory method here suggested lends itself to a determination of what elements of quality and service in a given product are deemed most essential by the consumer. The effectiveness of the ideas conveyed in building up a demand reflects the intensity of human wants as to the elements of quality and service described. The producer can sound the consumer and can better adapt his product to the consumer's felt needs.

Thus an entire selling campaign can be directed on the basis of what may be termed laboratory study. The empirical methods of the ordinary business man may be supplemented by scientific methods that have proven efficient in other fields.

The above practical suggestions have been directed primarily to the business man struggling with his immediate problems. Yet it may be well to emphasize once more the social importance of the suggestions. It is not merely that a large annual waste in advertising can be eliminated. Our whole system of distribution is in chaos. And the chaotic conditions in distribution mean that matter is ill adjusted in form and place to human wants. Only as systematic and widespread study along the lines indicated is given to the problems of distribution, can we build up an organized body of knowledge as to the facts and principles involved. And only on the basis of an organized body of knowledge about distribution can we hope to work out a more efficient organization of distribution.

And to this end the business man must coöperate with the scientist of the university. Much can be done by the trained student in his laboratory or in his study that will be of practical value in making possible a more efficient organization of distribution. The experimental psychologist can do much to work out general principles that will aid the business man in solving definite selling problems. The difficulty has been that the laboratory worker does not have the specific problems of the business man brought to his attention.

Similarly, the universities, through investigators trained in economics, can gather and correlate data upon distribution that will be of enormous practical value. They should, through research bureaus, study such problems as the cost of distribution in the various industries at different stages. And gradually a body of organized knowledge of the actual facts of business will arise. It is by development along such lines that future improvements in the system of distribution will be made possible.

REVIEW

1. What is the writer's idea of a market? Contrast market *area* and market *contour*.

2. How does the writer support the following thesis with respect to markets: "A clear grasp of the problem through a careful analysis is the first step in solving difficulties"?

3. What is the "laboratory method" in business analysis? What claim has it to be called "scientific"? Contrast it with that known as *a priori*.

4. Illustrate the application of the laboratory method to advertising. How does the law of averages in mass phenomena apply here? Is the case different in the determination of price policies? Why?

THE MEASUREMENT OF THE RATE OF FACTORY OUTPUT[1]

Enumeration

The enumeration of any type of output depends upon its *uniformity* and its *divisibility into units.*

The first task for every investigator proposing to use output as a measure of working capacity is to find *uniform* operations performed throughout the period to be studied. At a large munition factory an attempted comparison of the different week's output of certain girls nominally on the same work was made impossible in the majority of cases owing to the fact that the girls were not really continuously on the same operation. One week a particular girl working on a capstan lathe was set to make one part of a fuse, in another week or even in the same week she was making another part, of quite different complexity, and, therefore, with a quite different rate of output per hour. Indeed over the whole factory it

[1] Adapted with permission from Florence, Philip S., "Use of Factory Statistics in the Investigation of Industrial Fatigue," *Studies in History, Economics and Public Law*, Columbia University, Vol. LXXXI, No. 3, 1918, pp. 39–55.

was only in one 18-pound shell cartridge case department and in the work of six girls in the fuse department that the kind of output was found sufficiently uniform over a long period for purposes of enumeration.

The investigator should be especially on his guard that products known by the same name are not of slightly different size, or for some other reason do not vary in the effort required to make them. The output of an individual may be recorded on paper, as so many unit "boxes," but when the matter is investigated the actual output will be found to fall into various amounts of say 2-ounce boxes, 3-ounce boxes, 4-ounce boxes, with no common measure of the respective requirements of each in the amount of activity exerted.

Where the output is thus of various kinds, a sort of common denominator may sometimes be found for all the varieties in the amount of piece wages earned, or where the task bonus system has been introduced in the degree of efficiency attained. The accuracy of this denominator would depend of course on whether the piece rate or percentage efficiency was estimated exactly proportionately to the comparative effort required of the worker as between different varieties of output. My own experience with the measurement of working capacity by piece rates and by efficiencies, even where these had been estimated by the most careful time and motion study, was unfavorable to the use of such common denominators. In one factory that I visited the amount of task bonus paid for many processes depended on the percentage of efficiency attained, and much trouble was taken to insure that 100 per cent efficiency in each variety of work entailed exactly similar effort on the part of the worker. Now in many departments a great fall had been taking place in the efficiency attained. But it was admitted by representatives of the firm itself that this fall was probably due merely to a change from one kind of work to another. At my request a study of this factor was made in one department, and there it was seen that "efficiency" clearly varied according to the variety of work being performed. It seemed impossible to compare numerically the degree of effort required in different work.

This difficulty, of course, in no way nullifies the calculation by piece rate earnings or by efficiencies where the same kind of output is being produced throughout. If the record of earnings and efficiency is more accessible, by all means let it take the place of the direct output record. . . .

Comparisons of the cost of labor as a common denominator for all varieties of work will give a still rougher measure of working capacity. It does not avoid the discrepancy between comparative piece rates and comparative effort, and in addition raises discrepancies in the actual computation of the cost.

If a worker is employed on different operations it may be possible to select for comparison the output rate of any one operation that recurs regularly at intervals. The difficulty here, however, is that the output rate of the operation that is selected will be affected by the degree of effort required on the various operations preceding it; and at each recurrence of the operation studied, the preceding operations may be different.

Operations that result in a quantity of *units* being produced are confined to what the manufacturers and workers usually call repetition work. How many such units must for statistical purposes be produced per day depends on the period studied. If the hourly output is being compared the repetition must obviously be more frequent than if only the daily output is the subject of comparison. To show variations as between different periods with any exactness at least three units should be produced on the average in each period compared. Sometimes the timing of output is given not as the number of units per hour or per day but as the number of minutes or hours per unit. This, however, is easily translated to units per period and the same rules as to frequency apply.

Luckily for the investigator, though possibly not for the workers themselves, such frequently repeated work has been increasing under the modern factory system owing to the continual replacement of men by machines and the continual division of labor. Work is stereotyped and work is clearly defined. This applies very particularly to the munitions industry where products have to be made according to government "specifications." The munitions industry accordingly supplies a very fine field for output records.

Appended is a list of a few processes producing enumerable units that are sufficiently repetitive to have been used either by the present writer or by fellow-investigators as measures of working capacity.

Packing Processes.

Straightening rods or cans with a hammer.
Sticking labels on standard-sized cans.
Soldering lids on standard-sized cans.
Filling standard boxes with products.

Assembling Processes.

Assembling links into a chain.
Assembling the fuse of a shell.
Covering middles (*i.e.* creams) with chocolate.
Joining sides and bottom of standard-sized boxes.

"Working-up" Materials. ("Machining" Processes.)

Sewing belts and buttonholing by machine.
Drilling, boring, etc., parts of shell-fuse.

Lathe-work on standard 18-pound shells or any standard "parts" of a fuse.

Machine-tending (semi-automatic).

Feeding machine with cartridge cases.

Feeding, emptying, and controlling presses.

Typesetting by hand on typograph.

The same processes or crafts are of course often found in different industries. The munition industry, for instance, includes many processes found in automobile manufacture.

Expressiveness

Once a type of output is found consisting of a number of units which can be said to vary "up or down" because it consists of a greater or lesser number of units, the next stage is to select such an enumerable kind of output that these variations will be expressive of variations in the degree of working capacity. In the case of measurement by output such expression, if it exists at all, will of course be "congruent," *i.e.* when working capacity increases the output rates will increase also and *vice versa*. . . .

Elimination of Ambiguity

To enable the rate of output to measure working capacity without ambiguity the influence of factors in the industrial situation must be excluded that modify output one way or the other without passing through "capacity" first. Factors likely so to modify output must be kept "constant," so that changes in output cannot possibly be attributed to any changes in these factors foreign to our study.

If, for instance, the output of a factory was falling from one week to another and hours of activity had been raised, it would not be possible to *prove* that the decrease of output had measured a diminution of working capacity unless it were certain that the type of workers and all other factory conditions had remained constant. Otherwise the fall in output might just as well be attributed to a more inexperienced set of hands.

The chief factors that are likely by their inconstancy to disturb or make ambiguous the relation of output and working capacity are connected first with the type of worker and secondly with certain working conditions. They comprise:

A. The Type of Worker.

B. The Preparedness for Work.

C. The Stimulus to Work.

D. The Feasibility of Work.

A. *Constancy in Type of Worker:* Where a whole factory's output is under observation it is obvious that the total may quite likely be the product of an ever-changing set of individuals or even of an increasing or decreasing number.

Where the number is changing the total output should be divided by the numbers employed and expressed as a rate per individual worker. Sometimes the actual number at work cannot be, or at any rate has not been, ascertained. Though the number of machines or work benches is known, yet a few workers may have stayed away all day. In one munition factory I found records giving the total output per shift in each process, irrespective of the number of individual girls working at the time. But as it was to the interest of the management to keep every one of the machines at work, a reserve of girls was kept to be put to work in case the girls usually employed did not appear. Hence it is not likely that the number actually working varied much as between the dayshift and the nightshift on the same date. Mass statistics such as these, though inexact when taken alone, are often useful for checking the results of intensive studies.

Even when known, the rate of output per individual is likely to diverge from working capacity if the employees as a whole vary in their skill or experience.

A comparison attempted by the writer in a munition factory between the output rates of girls working two eight-hour shifts and girls working one twelve-hour shift had to be abandoned because the number of girls employed on the one shift was only half that on the newly instituted two-shift system, hence every second girl in the short shifts had been freshly hired and was inexperienced. The average output for the short shift was lower, therefore, not because working capacity had diminished among certain given human organisms but because organisms of a lower capacity had been added.

Again, at another munition factory hours had been increased in the first year of the war and efficiency had fallen, but the latter was not with any certainty attributable to a diminished working capacity in the same individuals. Besides the increase in the hours of work there was a constant increase in the number of new hands taken on. In one department a great number left to form a new fuse-making department, and their places had to be filled by new workers.

It is clear enough from this discussion that the only factory records really free from ambiguity are those specifying the output of each individual worker. The investigator should always endeavor to *compare only similar work from the same worker or group of workers.*

B. Constant Preparedness: Even when the type of worker is constant, or when the output of exactly the same workers is studied throughout, certain working conditions are liable by their inconstancy to render the output an ambiguous measure of capacity.

First of all, conditions may not always be *ready* for work to take place. Working time may be wasted and not "filled in" with work. The worker may be waiting

(1) for his material to be brought to him or
(2) for his machine to be repaired or
(3) for power to be connected with his machine.

Conversely, material, machine, and power may be waiting for the worker. He may be late coming in or late getting ready and preparing his materials, or he may be called away for payment of wages or duties about the factory or he may be allowed to leave early at the end of the day or start his tidying-up early.

All these cases of stoppage or tardiness may be considered involuntary waste of time, in the sense that the work did not take place, because physically speaking it could not be performed; the worker and his equipment were not *prepared* for the task.

In his table of output the investigator must note separately the time that was thus wasted involuntarily, and that wasted willingly, as in talking, resting, eating, voluntarily leaving room, etc. Allowance should only be made for the time lost *in*voluntarily. The investigator must consider all the hours and minutes the worker actually was ready to work, and only those, and base his rate of output on that as denominator, *e.g.* if the worker was prepared only for 40 minutes of the hour his output *rate per hour* should be his actual output multiplied by 60/40. The output is "corrected" in the same proportion as the nominal time was to actual time prepared for work. Thus, where output is reckoned up hourly the table might run somewhat as follows:

Hour	Gross Output	Time Wasted Involuntarily	Corrected Output	Time Wasted Willingly
9–10	20 Boxes	9:30–35 Machine Stoppage	$20 \times \frac{60}{55} = 21\frac{9}{11}$	Rest 9:10–9:20
10–11	15 Boxes	10:40–11 Lack of Materials	$15 \times \frac{60}{40} = 22\frac{1}{2}$	Leave Room 10:20–10:25
11–12	12 Boxes	Call to Office at 11:40	$12 \times \frac{60}{40} = 18$	Talk 11:30–11:35

The length of the stoppages due to late arrival or early quitting of the workers may be discovered in most factories by an automatic clock which stamps the exact time on a card inserted by each worker as he enters or leaves. These "clocking in" and "clocking out" cards are then usually taken to the wage office.

Stoppages in the course of work can usually only be noted by direct observation. Either the foreman or the investigator himself must be prepared to time any stoppages of more than three minutes' duration.

C. Constancy of Stimulus: Now, even where industry is as regularized as it is in the factory, there are many motives playing upon the worker that vary in force from time to time. The worker during working hours must not only be constantly ready and prepared to work but he must be constantly willing and eager to work as well. The investigator must make certain that workers are not discouraged nor "sulking," nor yet controlling their output deliberately.

In one highly organized munition factory records taken by the firm itself on drilling work showed that "the rate of production drops heavily whenever the girl loses confidence in the accuracy of her work." Conversely, "a stoppage due to breakdown, if repaired so as to give the girl confidence, causes an increase of speed." One of the explanations offered for this, namely the desire to make up for the stoppage by faster work, is paralleled by the haste often exhibited at the end of a working period in order to finish off a given operation or complete a given task. All these are cases where the stimulus is inconstant owing to variable moods, and the disturbing factor can be exercised by averaging out.

On the other hand, a very striking instance of the stimulus being regularly inconstant owing to deliberate calculation was discovered at a large English munition factory. A certain definite amount had apparently become the traditional day's output. If the worker approached this output earlier in the day than was usual he would usually slow down deliberately to avoid "exceeding the limit." To detect such limitation of output that is not necessarily due to diminished working capacity, the investigator should look back over the records. The stereotyped repetition of exactly the same number of units of output by one worker after another, week after week, is highly suspicious.

In certain cases the incentive to work varies owing to the stress of economic circumstances upon the business pursued. Work in offices, for instance, is subject to special rush hours during the day when the mail must be dispatched. Such diverse industries as laundries and telephone exchanges are subject also to rush days during the week, or rush hours during the day, when the demands of their customers are heaviest.

During these times the factory or office management will incite its staff to special efforts and any slackening will lead more readily to dismissal than at other times. As a result, output will rise during the rush. In the office of a munition factory, for instance, a typist working from a dictaphone was found to average anything from 2.16 to 3.83 lines a minute from 5 to 5:45 P.M., when dictaphone records had to be immediately transcribed into letters, but her average at other times was about two lines. This did not mean that her working capacity was greater at 5 in the evening but probably that the same capacity was stimulated to greater efforts.

The constant desire to earn high wages can be relied upon as an incentive to work to full capacity, and an incentive strong enough to overcome all the other various motives, only when such wages are paid on a piece basis; that is to say, when the amount of earnings depends on the amount of work done. Investigators are strongly advised not to make records of outputs under a time-wage system or even under a piece-wage system that is strongly digressive [*sic*] (where the greater the output the less in proportion is paid in wages) unless discipline and the fear of losing employment and all wages are unusually potent.

Above all, output produced under different *scales* of wages should never be compared. Overtime work, for instance, is often paid at one and a quarter or one and a half times the piece rate paid for work during the normal working day and extra work on Sundays is often paid double. As a result workers will tend to "go easy" in ordinary hours or on week-days and reserve their strength for the overtime and the Sunday work. Output will vary accordingly but it will furnish no clear indication of working capacity.

A similar variation in what is after all the main incentive in modern industry, namely the "economic" motive, may sometimes be found owing to the maladjustment of different wage systems. In a small munition factory near London, though piece wages were nominally being paid both on an eight-hour and a twelve-hour shift, girls working the short shift were in certain processes being remunerated *in fact* only by a time-wage, since they knew, or thought they knew, beforehand that they could not produce enough output in the shorter hours to earn more than the minimum hourly time-wage which was guaranteed them by a trade-union agreement. On the long shift, therefore, girls were likely to be "trying" much harder than on the short shift.

When the main incentive is not a constant force output data are rendered useless. The degree of inconstancy cannot be measured accurately and the investigator is warned never to choose records under such conditions.

D. Constancy in Feasibility: To measure working capacity unambiguously, variations in output must obviously not be due to variations in such foreign circumstances as the quality of the materials and of the machines used in the work or to the quality of the lighting.

Lighting, besides influencing output indirectly through its influence on working capacity, particularly that of the eyes, may affect the ease of operation directly and physically by its influence on the visibility of the material equipment. The Industrial Commission of Wisconsin found that a certain steel plant by merely changing its system of lighting increased its output at night by over 10 per cent, and undoubtedly any excess of output by day over that at night is in part attributable to the greater power and more equal distribution of daylight. In certain processes, however, artificial light can more easily be centered on the work and glare can be avoided.

The same amount of a given kind of output if produced from different machines may have involved quite a different ease of production; and even similar machines will vary substantially in ease of production according as they are oiled, connected with the power, etc. The investigator should hesitate, therefore, before classing as identical even similarly named and similar-looking machines. The slightest difference, when the machine is at work, in the methods of driving, feeding, and controlling it, and guiding the material will produce vast differences in the feasibility of a given operation.

Raw material, even of exactly the same name, when drawn from different parts of the globe is likely to differ greatly in the ease with which it can be handled — in its softness, malleability, pliability, etc. Again, it is well known that cotton thread while being spun breaks less easily in a humid than in a dry atmosphere.

The quality of the raw material supplied may vary also according to the skill of the operator who prepared it. Thus in cotton-spinning, the number of threads that break on the slobbing frame depends largely on the skill displayed in the drawing processes that just precede the slobbing.

Because of the enormous differences in feasibility of any given output due merely to differences in factory equipment and technique — lubrication, lighting, materials, machines, and also to factory organization — *it is inadvisable for any investigator to attempt to compare the working capacity in one factory directly with that in another.*

Sources of Record

The method of collecting output data that is most likely to be accurate is for the investigator himself to watch a group of workers and note their output, staying in the factory day in and day out, and this method has the advantage of continually suggesting to the investigator new facts of significance and new methods of recording them. For instance, as I watched the output of four girls assembling bicycle chains with a press driven by foot, for two days of eleven hours each, I observed clandestine meals and rests taken unofficially and how the rests were spent. Further, struck by the constant rhythm of the girls' motions, I was led to some new investigations into the value of rhythm as a stimulus.

However, the personal collection of sufficient output data to establish conclusions would require a whole army of investigators, and even then the presence of the investigator is only too likely to disturb and make unrepresentative the very facts he wishes to secure in their native state, as actualities of industrial life. Indeed I found that the average output of the four chain-assemblers was at a speed considerably higher than usual on the days I watched, being 7.10 chains per hour as against from 5.85 to 6.80 recorded in the books for previous weeks. In spite of a tactful explanation of my purely scientific purpose, the presence of a stranger making strange notes may have inspired a fear of taking very long rest pauses or of indulging too much in conversation. Where, however, as in "scientifically managed" factories, the workers are accustomed to being time-studied, the disturbance due to this factor will be much smaller.

The method of recording output which is least disturbing to the worker's ordinary attitude and also most easily carried out is by use of automatic registers, of which the cyclometer is perhaps the most familiar type. I have seen clocks or registers attached to machines such as looms, stamping-presses, sewing-machines, where each revolution of the crank producing a unit of output was duly recorded in figures which could be read off whenever required. Some registers are even self-recording; that is to say, instead of being "read off" by human agency they actuate a pen which traces the curve of output on a rotating drum. In view of the low cost of registers and the ease with which they are attached, their use might well be extended.

A method of recording only slightly more disturbing is for a member of the factory staff, usually the foreman, personally to make the record. To the worker the presence of the foreman and his taking of notes are a part of the factory routine;

the worker's attitude will not alter much from that of his ordinary working mood.

Such records either personally or automatically collected by the firm may either be initiated by the investigator or may already be in existence when he begins investigating. As the investigator enters the factory for the first time, somewhat bewildered perhaps, he should ask that the output records already collected be shown him. Never should he lose an opportunity of using the documents of industry that he finds ready to hand. He cannot be urged too strongly, however, always to subject these factory records to a detailed scrutiny. First of all, he should visit the actual operation in the workshop of which the record shows the output. This personal visit, especially if the investigator has even a small knowledge of mechanics, will probably suggest explanations of peculiarities in the records or perhaps show up errors in recording. Secondly, the output record itself should be carefully checked and the same questions put as though the investigator was selecting operations to study for himself. Was the output enumerable? Was it expressive? Was it free from ambiguity, with personnel, preparedness, incentive, and feasibility either constant or averaged out? Records when kept in the factory books as a matter of routine often range over a long period and cover a large number. As mass statistics, therefore, they will offer a great chance of averaging out inconstant factors and, even when not entirely free from ambiguity, they may often prove useful in checking intensive inquiries. I have used figures of gross output per machine, irrespective of possible absences of the workers, in a whole department making millions of rifle cartridge-cases per week, to check a comparison of night and day efficiencies based on the weekly output records of *selected* individuals. It was very much against the interest of the firm to have any machines lying idle, so that absences, or at any rate absences without substitution of another worker, were extremely rare.

Seldom, of course, will these records of output have been made by the firm for the purpose of studying working capacity; when they are taken it usually is for the purpose of computing the piece-wages to be paid their workers.

In one munition factory where workers are paid so much per thousand rifle cartridges turned out, with a minimum guaranteed wage of so much per hour, the hours worked and the output on each day are noted down quite simply for each individual in small memorandum books kept by the foreman, and hours and output are added up for the week.

In another and larger firm, where the wage paid is based on a more complicated system and where the output is more varied, a huge "detail sheet" is kept at the "wages office" and filled in for each individual worker each week, being arranged as follows: columns are provided for the time at which the employee entered and left the works; for the time lost and the time worked for each day of the week. Each of the different kinds of operation the employee has performed is then entered item by item down a column; and opposite each entry is stated the hours worked on that operation and the output, both hours and output appearing under the proper day. Beyond the columns for each day are columns for the hours worked and the output of each operation for the whole week.

These columns contain the whole of the information on the facts of output rates that we require; columns beyond them work out the wages payable for the week from the facts already given.

REVIEW

1. What is the denominator; the numerator; in the coefficient "Output Rate"? What measures may be used to determine it, or to reduce it to a common denominator? What are the limitations of each?

2. If the aim is to measure statistically factory output, what conditions may occur respecting

(1) The type of worker.
(2) The preparedness for work.
(3) The stimulus to work.
(4) The feasibility.

Which will make the result ambiguous or "indeterminate"? Make a list of the things under each heading as given in the text, and add others from your own experience.

3. Who should take the record of output? Why? What tests should be applied to determine the use and value of records? Make a list of them and compare them with those given in the *Text.*

4. Is the above discussion, in relation to methods and safeguards in collecting statistical data, of universal application? If so, show how they apply in such problems as

(1) Studying wage data as a basis for an arbitration proceeding.
(2) Studying accidents as a basis for introducing safety devices.
(3) Analyzing sales as a basis for an advertising campaign.

WHAT'S IN A NAME — THE CAUSE OF DEATH [1]

Error in the Official Record of Deaths from Tuberculosis. — There can be no doubt that the tuberculosis rate was diminished by inaccurate statement of the cause of death on the official certificate. In a large number of cases the cause of death certified to by the physician was contradicted by the history of the decedent's illness as reported by relatives.

[1] Adapted with permission from "Errors in Death Registration in the Industrial Population of Fall River, Mass.," in *Monthly Review*, United States Bureau of Labor Statistics, Vol. 5, No. 1, July, 1917, pp. 2–8.

Thus, in cases in which the physician's certificate gave some such equivocal cause of death as bronchitis or hemorrhage, or some terminal conditions, such as broncho-pneumonia or heart failure or debility, relatives of the decedent testified that for possibly a year or more before death the decedent had had a bad cough, had expectorated profusely, had become extremely emaciated, had suffered from night sweats, had had one or several hemorrhages of bright blood, and was the second or third in the family who had "died of consumption" within the last few years, or had parents one or both of whom had died long ago after years of such tuberculous manifestations. Such testimony as to matters of simple fact seems entitled to considerable credence.

A French-Canadian woman, aged 23 years, . . . for 7 years a spinner until she left the mill because of cough two years before death, was certified by her attending physician (now dead) as having died from "bronchitis." Another attending physician whose name is upon death certificates of two other family members did not "recall" this case. The seemingly tuberculous mother and brother of decedent affirmed that the latter had died from tuberculosis, "just as her father and three sisters did." These last mentioned four are certified as having died of tuberculosis between March, 1910, and August, 1912, and are so recorded in this study. Another sister was recommended to a tuberculosis hospital October, 1909, and is said to have recovered. This case was scheduled as nontuberculous. . . .

A special canvass was made to see just how commonly tuberculosis was misreported on the official death certificate. There were 188 causes in which there was marked discrepancy between the cause of death as given on the death certificate, and the cause of death suggested by the history of the decedent's illness as given by the family. Every physician who had signed one of these 188 certificates, if still living and still in Fall River, was visited and questioned about the death. By this process the probable correctness of the certified cause was satisfactorily established concerning 31 of these cases.

In 65 of the remaining 157 cases no further information was obtainable, because the certifying physician had either died or left Fall River, or else professed inability to remember and no other attending physician could be found. In not a few of these 65 cases the histories indicated overwhelmingly that these deaths were due to tuberculosis. Nevertheless, the certificates were taken as correct unless an admission was secured from the certifying physician that the recorded cause of death was incorrect. Consequently these 65 cases have been counted as correctly certified.

The remaining 92 cases are either admittedly or demonstrably cases of tuberculous deaths. . . . These 92 cases may be divided into the following classes:

1. Those in which the certifying physician unequivocally stated the cause of death to be tuberculosis. These numbered 70.

2. Those unequivocally vouched for as tuberculous by a physician who had attended the decedent in his last illness but had not signed the death certificate. Recourse was had to these other physicians only because in every one of these cases the physician who had signed the certificate had either died, left Fall River, or forgotten all about the case. This forgetfulness is explained by the fact that the signers of the certificate were sometimes city physicians, who had responded to an emergency call and possibly had seen the decedent professionally only once. These cases numbered 12.

3. Those who, after a sputum examination, had been recorded on city or hospital records as tuberculous. Of these there were five.

4. Those stated by the certifying physician to have been "tuberculous probably." Two of these had not been certified as tuberculous, because no bacteriological examinations of the sputum had been made, "and so," said the physician concerning one of these, "though I knew the case was tuberculosis I couldn't actually swear it was." This group likewise numbered five.

As a result of this special canvass, it appears that not improbably one-sixth (17 per cent) of all the fatal tuberculosis in the city was misreported under nontuberculous diagnoses.

Reasons for Erroneous Certifications of Death

The question of course arises why the true cause should be so often ignored or misleadingly reported on the death certificate. There seem to be several reasons for this. Some persons are sensitive as to the existence of a case of tuberculosis in their family and would seriously object to having such a cause recorded upon a certificate. The knowledge that this feeling is common may affect the physician even in cases where no such prejudice exists. But apparently by far the most effective reason is the attitude of some of the insurance companies which may delay payment of policies of decedents officially certified as having died from tuberculosis and which also not uncommonly refuse to insure other members of the family of such a decedent. Physicians when asked about these variant cases occasionally admitted that the certificates were designedly misleading, but justified them on the ground of personal financial expediency arising from intense medical competition, and on the added ground that sometimes only through such registration practices could the decedent's family secure promptly the amount they were entitled to from the insurance companies.

Error in Official Record of Decedent's Occupation. — In addition to the errors concerning the causes of death, whether principal or contributory, the records were found to be seriously inaccurate in their statements concerning the decedent's occupation. Fortunately it was possible to correct these errors to a very considerable degree, far more so than to correct errors in the alleged causes of death. As stated above, the physician's official statement as to the cause of death was accepted unless the original certification was admittedly or evidently wrong. This policy was followed no matter how seriously the correctness of the certificate was doubted. But a similar adherence to the record was not considered necessary in regard to the statement of the decedent's occupation, this being a matter on which the physician's professional training would have no bearing, and of which neither he nor the hurried and sometimes careless undertaker probably had personal knowledge. When, therefore, a statement by relatives or friends as to the occupation of a given decedent differed from that of the death certificate, the former was taken as authoritative.

The errors of the death certificates as to occupation were both of omission and commission. Persons who were really cotton-mill operatives were not so recorded. Others were set down as operatives who had never worked in a cotton mill or who had not done so for more than two years preceding death.[1] The former error was the more common among female and the latter among male decedents.

The extent of these errors as accurately determined in Fall River for the whole eight-year period — 1905 to 1912 — shows most conclusively the seriousness of the misapprehen-

[1] A considerable part of this error is due to the vague use of the term "operative," which is frequently employed on death certificates with nothing to show whether the person concerned worked in cotton or woolen mills, in dye works, bleacheries, or printeries, or in piano or hat factories.

sion which would be caused by using the official certificates without investigation of their accuracy.

For the eight-year period nearly one-half (49 per cent) of the female decedents who were found to have been cotton-mill operatives were not so recorded. On the other hand one-eighth (13 per cent) of the females recorded as operatives were found on investigation not to have been cotton-mill operatives. Among the males for the same period, 23 per cent of those who were finally classed as cotton operatives were recorded on the death certificates as following some other occupation, while one-fourth of those recorded as operatives could not properly be included among cotton-mill workers.

The recorded number of male operative decedents in Fall River for the eight-year period (1905–1912) was 915. Of these 233, or 25 per cent, were found not to have been cotton-mill operatives, while 207, who on their death certificates were assigned to other occupations, were really cotton-mill operatives at the time of their death. The real number of male operative decedents, therefore, was 889, the group as recorded having been larger by 26 than the facts justified.

On the other hand, the recorded number of female operative decedents in Fall River for the eight-year period was 548. Of these 71, or 13 per cent, were found not to have been cotton-mill operatives, while 459, who were recorded either as having other occupations or no occupation at all, proved on investigation to have been really cotton-mill operatives. This gives a total of 936 decedent female operatives.

Conclusions

There is no reason to suppose that the official registration of deaths is more carelessly or recklessly performed in Fall River than elsewhere; indeed, in view of the advanced position which Massachusetts has taken in regard to vital statistics there are grounds for the opposite supposition. It is believed, therefore, that the facts disclosed in this summary show:

(1) That there is urgent need for a closer supervision of death registration, and for a sustained effort to secure greater accuracy and a nearer approach to completeness in the certificates filed.

(2) That a small minority of the physicians of a city or State are able most seriously to retard progress in industrial hygiene and preventive medicine, through their failure, admittedly sometimes intentional, to give intelligent compliance with the death registration requirements of the law.

(3) That under present conditions death certificates need careful verification before any but the most general conclusions respecting early death in industry may be safely drawn from them. In particular, deductions as to the prevalence, the increase, or the decrease within any specified age group of fatalities from causes like tuberculosis . . . or as to the effect of a given occupation upon those of a designated age who follow it, are liable to be wide of the truth if based upon the death data of the registrar's office, unless such data are first subjected to detailed investigation. . . .

REVIEW

1. Is the inaccuracy in the cause of death due to reporting or to the determination of the cause?

2. What were the causes for the errors in the occupations? What is an "operative"? Formulate a definition which can be statistically used.

3. Put into a brief statistical table the numerical facts contained in the last two paragraphs previous to the *Conclusion*. Does the tabular form help to make the figures "stand out"?

Statistical Standards in the Collection of Facts[1]

First, facts must be collected for a definite purpose. Statistical analysis cannot proceed as if it were in a vacuum; the meaning of a statistical fact is a function of the use to which it is put, and the costs of collection are justified only in the realization of a purpose. For collection to proceed without a definite goal in mind is not only wasteful of time and money, but fatal to the idea of statistical analysis. Facts are not equally good for all purposes. The acts of measurement and of classification presuppose a purpose. Fruitless investigations carried on at enormous costs and resulting in ill-will on the part of those who are interested in the results, discouragement on the part of those who are undertaking them, and a tendency to scout the idea of statistical analysis and the function of experts, are largely if not solely traceable to a violation, of this seemingly self-evident truth.

Second, facts must be collected in standardized units and under uniform methods of application.

Third, a sufficient sanction for the collection or use of data must be secured. To formulate a definite purpose for which facts are wanted is the first condition for securing this. It is generally, but not always, necessary to demonstrate that personal advantage will result from a study of the facts furnished. But often more than this narrow appeal may be made. Interest in *fundamental principles* may be aroused.

Fourth, standards of collection require that the full import of such questions as the following shall be considered. (1) For what periods, under what conditions, and for what places are the facts available? Are the purposes and methods of analysis

[1] Adapted from Secrist, Horace, "Statistical Standards in Business Research," *Quarterly Publications of the American Statistical Association,* March, 1920, pp. 51–53.

conditioned by the answers secured? (2) Will *available* facts be given or may they be assembled; and if so in what units, with what degree of accuracy, and with what effect? (3) Do the schedules or forms used in collection provide for keeping confidential the data supplied? (4) Are the units of measurement which are employed standardized and understood? Do they follow or run counterwise to the terminology of the records employed? How may necessary adjustments be made and with what effect?

Fifth, statistical standards require that wherever possible the truth or error of facts shall be verified. Against the imputation of gullibility, those in charge of statistical analysis should always be capable of defending themselves. To take on faith the plausible or to discard seeming exceptions is not in keeping with scientific method. Verification requires more than testing mechanical accuracy and removing apparent inconsistencies. It involves an analysis of the composition of groups and totals, and a scrutiny of the uniformity of measurement and the methods in which units are applied for different times, places, and conditions.

Sixth, the field from which data are secured must be adequate and the facts inclusive or representative. The choice of the field and the selection of the facts depend upon the purpose for which analysis is undertaken. A problem requiring inclusive data must be approached differently from one which may be studied by means of samples. Standards of collection, may, indeed, become standards of elimination; and balance and consistency rather than simple verification of accuracy become the goal.

CHAPTER III

UNITS OF MEASUREMENTS IN STATISTICAL STUDIES

The Nature and Conditions of Statistical Measurement [1]

It is very seldom that the unit, which is actually used in compilation, is that which the unwary would imagine from a carelessly quoted summary, or that which *a priori* an investigator would desire. What constitutes a pauper? What entitles a man to be included in the Labor Department's monthly total of unemployed? What relation have the totals of paupers or unemployed officially stated to the poor or unfortunate as to whose condition and numbers we desire knowledge? What does the Labor Department understand by an increase of wages? What is income, and what relation has it to the total income published by the Inland Revenue Commissioners, determined by numerous Acts of Parliament and limited by judicial decisions? Under what circumstances are married persons returned as unmarried, or *vice versa*, for census purposes? What is a room and what a tenement? How is the value of wool and how the value of machinery or of pictures determined for the trade accounts? Does the total value generally quoted for our foreign trade include bullion, ships' stores, ships' coal, ships themselves, foreign produce bought and re-sold, or transhipped in bond, and given

[1] Adapted with permission from Bowley, A. L., "The Improvement of Official Statistics," in the *Journal of the Royal Statistical Society*, September, 1908, Vol. 71, pp. 463–469.

the answers to these questions, how far must they be modified for other countries? The greatest of the difficulties in comparing the statistics of two countries is in obtaining adequate definitions of the units. The definition is a matter of conventional and very elaborate delimitation, sometimes arbitrary, sometimes dependent on law or on custom; and till the principles of the delimitation in each special case are known the statistics resulting cannot be used with any certainty.

Homogeneity. — It is frequently the case that the most distinctive attribute of the unit is variable in degree or not capable of exact definition, or that for other reasons the units, similar in the attributes selected, are dissimilar in other equally important attributes. Let us consider the contents of some well-known totals. The number of adult male wage earners in Great Britain and in other countries can be estimated, but the relation of these totals tells very little about the labor power of the nation, for the men in one country vary greatly in skill and energy, and the range of skill and energy differs from country to country. The amount of wages received and work done vary from man to man, and the totals are composed of units which are heterogeneous for all practical purposes. If we aim at greater homogeneity by counting only the skilled men, we find that "skilled" is a term not capable of exact definition. Again, if we take Mr. Booth's or Mr. Rowntree's estimates of the number of persons below a fixed standard of livelihood, we find at once that the distance they fall below that standard varies greatly, that the pressure of poverty depends on many moral qualities and accidents of situation, and is not simply a function of deficit of income, and consequently that these totals are not homogeneous in the connection for which they are generally used. Or if we consider the total value of the exports from the

United Kingdom, we find that items included are heterogeneous for all purposes except the balance of trade. £1000 worth of exports of any kind gives rise to a bill of exchange that will purchase a corresponding amount of foreign goods; but as regards the employment of home capital and home labor there is every possible variety, from the export of coal, entirely a home product, to the export of foreign goods which become entitled to be called British by the process of repacking, and further the relative shares of capital and labor in the production vary indefinitely. If we are considering the profits to be made by a foreign trade, as compared with home trade, for instance, the root of the protectionist controversy, we find that there is no necessary relation between the total trade and the amount of profit, and that the various parts of the export trade are probably extremely heterogeneous in this respect. Still more heterogeneous is the total obtained by adding imports to exports, a quantity which is changed by some £15,000,000 by the alteration of 1 *d.* per lb. in the price of raw cotton, without any corresponding alteration in any of the things we wish to measure; and if we further divide by the population to obtain the £24 of foreign trade per head of the population, which is given a conspicuous place in the Statistical Abstract, we have a sum of essentially unlike quantities divided by a quantity which is heterogeneous in itself and has dissimilar relations to the parts of the numerator, for in no sense are the various units of the population similarly interested in exports or imports. The height of absurdity is reached when amounts obtained in this way are compared country by country.

There are two methods by which such difficulties may in some cases be overcome. The first is by subdivision by qualities, the second by grading of quantities. If we are comparing the number of operatives in the cotton industry, now and at a former period, the totals are heterogeneous in sex and age, and the comparison is misleading, for the numbers of adults and children, men and women, have not changed in the same ratios; but if we subdivide by age and sex, we can get a fair basis of comparison, and still better if we can make a further classification by skill. Thus we should continue to combine attributes in our unit, till one unit is similar to another as regards the purpose for which the totals are to be used. A case in point is the birth rate as corrected by reference to the number of married or marriageable women within certain age limits, instead of to the population at large. If it can be ascertained that the various dissimilar parts bear the same relation to each other in both totals, *e.g.* if there had been no change in distribution of age or sex or skill in the cotton industry, the heterogeneous totals may be used. The second method applies where the attribute, which is principally to be considered, is susceptible of measurement, as age or wage or income. We could not then say that one population was twice another, but could group according to age, and compare the groups, representing them by curves or mathematical symbols; and similarly we could deal with adult male wage earners, giving not only their number, but their distribution by wages. It should be said that an average should always be suspected, till the extent of the homogeneity of its numerator has been tested, in relation to the purpose for which it is to be used.

We cannot in general obtain perfect similarity in our units, without such subdivisions as would leave us with a number of unrelated units, instead of with a statistical total or average; but we can often get sufficient similarity for practical purposes. The total number of persons relieved under the Poor Law on a given day conveys no useful information; but if we could get the number of persons of various ages and of various degrees of physical and mental capacity tabulated, we should be able to make useful comparisons from place to place and time to time.

Universality. — When an investigation is made it must deal impartially with the whole district, the whole class, or the whole period in question. The general method of attempting to secure universality is to count all that is practicable and ignore the rest, thus introducing an error of unknown magnitude in the result. Two alternative or corrective methods are possible and in some cases easy to practice. The first is to make a careful estimate of the maximum or minimum differences, which would be made in the total or average if the missing part were included. If nothing whatever is known as to the omissions, especially if their existence is not even suspected, of course no correction can be made, but this is not the general case. Passenger journeys on railways can be calculated by the number of tickets issued, together with an estimate for the number of journeys made by contract-ticket holders. In the population census an estimate can be made for the travelers and homeless on the census night. In a wage inquiry a superior estimate can be made for the wage earners not counted, and the greatest possible effect on the average can be calculated. For the national income maximum and minimum values could be (but have not been) estimated for the incomes of non-wage earners not liable to income tax. Such estimates of the residuum are sometimes difficult and often unconvincing, but nothing is gained by ignoring them.

The error introduced by such absence of universality is of the kind I have called "biased," that is to say all or most of the different parts omitted are likely to drag the average in the same direction. It is the obvious that are counted, and their very obviousness is an attribute that differentiates them. In a recent American Wage Inquiry (supplementary to the census of 1901), the biggest firms were selected. More generally an inquirer aims at "typical firms," and very frequently he is limited to the firms who are the least opposed to an investigation. In none of these cases will the true average, as it would be obtained from a universal inquiry, be obtained. The selection of the typical firm appears to be the most plausible method, but, to put the criticism briefly, the "mode" will be obtained instead of the arithmetical average, and these two are not in general identical.

This leads to the second method of obtaining universality, that is the method of samples. I have recently dealt with this at length, and need only emphasize the chief essential of collection, that *every unit in the district or class dealt with must have approximately the same chance of inclusion*, and that the selection must deliberately be made at random; compared with this rule the number of units contained in the selected sample is unimportant. The only test of the adequacy of the sample is the similarity of results obtained by random subdivision of the sample. To test the purity of London water needs the examination of only a few microscopic quantities; to estimate the earnings of outworkers in West Ham would need a very extended inquiry before the accidents of the individual samples were eliminated.

Stability. — In modern societies the totals and averages which are the subject of statistical measurement are seldom stationary; some fluctuate with extreme rapidity and irregularity, some have fairly regular periodic changes, some grow or decline slowly and steadily. In the first case, frequent measurements are necessary for the presentment of any adequate picture; for example, when dealing with prices in general, with the earnings of pieceworkers, or with meteorological statistics. In the second case, the measurements must cover a complete period, after the length and constancy of the period have been ascertained; for example, with pauperism, with unemployment, or earnings in a seasonal trade. In the third case, which applies to population statistics and birth-, marriage-, and death-rates, and others, occasional measurements are sufficient, and intermediate values can be interpolated. In all cases, the frequency of the measurements should depend on the stability of the total or average.

Comparability. — It has often been pointed out that isolated statistical totals are nearly valueless, and that we need generally to study change or differences; that is, we need to measure similar totals differing in place or time. When the difference is in place, as, for example, when we compare working-class expenditures in Glasgow and in London, the analysis already given as to homogeneity and stability applies, but the homogeneity must extend over all the averages compared; the averages must be estimated by unchanged methods and like can only be compared with like. Under this test nearly all the comparisons that have been made between the standards of living in different countries break down.

When the comparison is made between similar totals at different dates, the rules are obvious and simple, but none the less neglected. The definition of the unit must be absolutely unchanged, and in this definition I have included the method of collection. The mistakes and omissions, all the "biased" errors, must be repeated. Like can only be compared with like. This is a hard saying, and seems to rule out all progress and all the improvements which form the subject of this paper. The remedy is to make changes infrequently, but permanently, and when a change is made, to collect the information both by the unreformed and by the reformed method, to choose the former for comparisons with the past, and the latter for comparisons with the future. In the simple cases, where improvement is made by simple extension, as in the case of the inclusion of the value of ships sold as exports, the comparison is simple, if the alteration is clearly made. Where improvements of tabulation are made, there should generally be little difficulty in the double tabulation necessitated.

Relativity. — I am using this word for the logical relation of two numbers which are brought together as numerator and denominator, or as factors. While comparability concerns the relation of like to like, relativity concerns the relation of one group of phenomena to a dissimilar group. Thus the quotients already mentioned, value of exports divided by population and number of births divided by number of wives, are cases of relativity. An example of a different kind is income corrected for the change in purchasing power of money. In order that a quotient, average, or rate may be perfectly valid, the numerator should be homogeneous and the denominator should be homogeneous, and each unit in the denominator should bear the same potential relation to the attributes of the units in the numerator. Thus the output of coal per hewer employed, the number of ton-miles per engine-hour, and the average earnings of self-acting mule minders are valid in this sense. The work of all hewers is to win coal, the purpose of the engine's motion is to drag loads, and all mule minders are engaged on similar machines and paid on a similar basis. The rigidity of the rule is not, however, necessary. Heterogeneity that leads to unbiased errors is admissible from the principles of averages, and when two such averages or rates are compared the denominator may bear any constant ratio to the ideal denominator. Thus if the relative number of hewers to all employed in or about a coal mine is unchanged, we may *compare* the outputs per head of all employed, instead of per hewer, without error. The consideration of relativity has, to take a well-known example, led to the "correcting factor" for urban death-rates; and it is because of the possibilities of error indicated that such care is necessary in interpreting income or wages in the light of index numbers based on wholesale prices.

Accuracy. — It may be granted that no statistical measurement satisfies perfectly the conditions now laid down. Any breach of these conditions leads to inaccuracy of result, in the sense that the total or average or other result obtained is not a perfect measure of the group as defined for investigation, and is a still less perfect measure of the group characteristic which we ultimately wish to know. The main thing to recognize in connection with official statistics is that their accuracy, in spite of the caution and systematic verification used in their computation, is only superficial. Their universality is limited by their methods of collection. The number of births, the income liable to income tax, the total value of imports, are not known if births are unregistered, income concealed, or diamonds imported in passengers' pockets. The measurements are not closely fitted to the quantities of which we want knowledge. We want to know the number of capable persons who cannot get work, and the value of net annual earnings in terms of the economic goods on which they are spent; the labor department returns do not profess to tell us the first, and it may be beyond the power of statistics to measure the second. Further, most statistics, official and others, fail in one or other of the respects discussed. The result is that statistical measurements are approximate, and should be frankly given with their limitations explicitly described, and with the maximum effects of their errors estimated. The supplementary inquiry, which such an estimate often demands, is very seldom made. In simple cases, where the measurements are rough, but the errors unbiassed, the numbers can be given accurately in round numbers; the population to the nearest ten thousand, say, average wages to the nearest half crown, the value of exports to the nearest £50,000,000, and so on. In any case we should avoid such a statement as "the number of illiterate persons above 10 years old in U. S. A. in 1900 was 6,180,069," where a very successful investigation could hardly get the hundred thousand correct, and the definition of illiteracy is vague, and also has very little relation to education.

We may now summarize the characteristics of good statistics. The unit of measurement should be absolutely defined, its attributes should be precisely those which are related to the inquiry, and the group should be sufficiently homogeneous for the purpose for which the measurement is needed. The collection should be actually universal or based on samples, scientifically chosen, with adequate tests of their sufficiency. A sufficient number of observations should be made to test stability. Only statistics collected and computed by the same methods and on the same definitions can be compared. When two unlike totals are brought into relation with each other, the causal connection between the units of the one and the units of the other should be close and inevitable. The accuracy of the measurement, as limited by the definition of the unit, should be calculable.

REVIEW

1. Contrast Professor Bowley's treatment of units of measurements with the discussion in the *Text*, Chapters II and III. In what particulars are they the same; in what way different?

2. What would you say to the statement that "Homogeneity is always relative; absolute homogeneity is unthinkable"? Is this true in the same degree for all problems? Illustrate.

3. Illustrate, out of your own experience, the significance in statistical study of Bowley's conception of "relativity."

4. Suppose you were asked to list all of the brick houses in a

certain section of your city; all of the female servants attached to the houses. What conditions would you set up for identification? Write the instructions to a group of clerks for such an enumeration. Would these instructions be equally good for all purposes? Why?

5. Consult the United States Census of Manufactures for 1914 for the definition of an "establishment." Compare this with the definition used by the Census for 1890. What is an immigrant? Where can you find out? What is a business failure (see Bradstreet's, January 31, 1920, p. 82)? Would you think it difficult to count such units? Why?

A Mile of Track [1]

It may seem that the mile of track is a kind of statistical unit that is very easy to deal with. Quite the contrary is true. Owing to the complicated character of the network of tracks of many companies crossing and in effect, through joint and often somewhat indeterminate rights of ownership, commingling with each other in New York City, resulting in frequent duplication or ambiguity of returns, and in the presence of a large amount of "special work" of all sorts, instead of there being almost exclusively straight rail, measurements and returns of track mileage furnish data that are about the most difficult to assemble and compile of any offered in this report, even apart from occasions for doubt as to how unused track is dealt with. Under these circumstances it is not surprising that some of the companies frequently remeasure their property and revise their figures.

REVIEW

1. Consult the secretary or some other official of the street railway in your city, relative to the meaning of a mile of track as used by the company.

[1] Adapted with permission from *Annual Report of the Public Service Commission of the First District of the State of New York*, 1913, Vol. II, p. 35.

2. What meaning does your city engineer assign to a mile of improved street? Discuss with him other possible meanings. Does he use both simple units and coefficients? What are they?

Accidents in Public Utility Statistics[1]

The value of any kind of statistics depends largely on the quality of the unit. In the casualty statistics here presented the units dealt with are cases of killing or of injury inflicted on persons, the agency being the street-railway companies of the city. In the broader sense, "injury" is properly the inclusive word, but it seems unavoidable to use it to mean less than fatal injuries. In the present report it is employed in this narrower sense.

At first glance, it might seem that there could be no question of the meaning of injury. A person killed is killed. A person thrown from a car and suffering a broken arm is injured. So far injuries are discrete and easily recognized units. But this is as far as the simplicity goes. In a collision, several persons may be mortally injured, but not killed outright. To classify as merely an injury a mishap that results in death within an hour is manifestly incorrect. On the other hand, if a person has a weak heart, and is severely shaken up and bruised and scared, as a result of which he dies of heart failure within a month, the cause of his death is primarily not the railway accident, but the physical weakness that existed independently of the accident. And yet mishaps may occur to people of normal or even exceptional health that undermine their health and strength and finally cause death directly traceable to the car accident, though not in point of time its immediate

[1] Adapted with permission from *Annual Report of the Public Service Commission for the First District of the State of New York*, 1913, Vol. II, pp. 137–140.

consequence. In a possible suit for damages it may be necessary to go deeply into the causes of a death. In addition to the question above indicated it might have to be decided whether the person was a suicide or not. In these statistical tables, however, we are concerned, not with the tragedy of each death, but with the numbers of deaths, and those numbers taken in connection with the volume and magnitude of the traffic. To take the number killed outright, add to them the number that happen to die upon the cars, and those injured who die at any time after the injury, would entail practically impossible labor in following up each case. In accident statistics we are concerned, not with the individual cases, but only with representative averages. From this point of view it is sufficient to draw the line between the killed and the injured upon the basis of a fixed interval of time occurring before death follows upon the accident. In the present statistics if death results at any time within three days, the case is counted among the killed; if later, it is classed as an injury, naturally or presumably a serious injury, though death may be so indirect a consequence and so long delayed that this classification is not certain. From the point of view of exact science, even with reference to statistical needs, the time interval in question should be so defined that the total number of deaths so classified in the statistics is the number directly caused by the accidents, but of course some occurring within an interval so fixed would be due primarily to causes other than the accidents, while a compensating number occurring later would be directly due to them. In fact, there are no means at present practicable for determining the proper interval in question thus exactly. But for purposes of statistical classification and comparison the interval may be quite arbitrarily fixed, and yet serve very well, provided the definition be clear and unmistakable.

At the other extreme, there is a similar difficulty of classification in drawing the line between what is an injury and what is not an injury. Laxity of definition at this point is to be expected, since much must depend upon the bare statement of the person most directly concerned, and he is not likely to be entirely disinterested in view of the possibility of his becoming the beneficiary of a damage claim.

In the sub-classification of injuries by kind, the difficulties of classification multiply. "Fractured skull" and "amputated limb" are definite enough or can be made so, but a "serious injury" not defined with the utmost care is of quite indeterminate significance. Probably the best method of defining with reference to seriousness, when the definition cannot be based on anatomical facts, is by way of the duration of disability. A hospital case is of course to be classed as serious, but a visit to a hospital for examination or observation should not be so counted. . . .

For purposes of statistical comparison with other years, and on occasion with other cities, it is necessary to reduce the absolute numbers for accidents to ratios. Since the movement of the cars causes most of the accidents, one very important ratio is casualties per car mile, or what is in effect the same and is somewhat more convenient with reference to the relative magnitude of the terms of the ratio, casualties per 100,000 car miles. In relation to injuries to passengers, the ratio to passengers carried is the better basis of comparison. A still better ratio, that is, to the passenger mile, is not available for street-railway statistics. The greater ratio of accidents to passengers on the steam railroads is of course largely explained away by the greater average length of ride of steam-railroad than of street-railway passengers. Moreover, this ratio of accidents per passenger mile — it may well be noted —

is probably subject to qualification with reference to the greater likelihood of accident per passenger mile at rush hours. The effects of such minor causes of possible misrepresentativeness, however, entirely disappear in most comparisons. The fundamental ratio for injuries to employees is casualties per given round number of employees. But since the employee is exposed to accident throughout the year, instead of for a fraction of an hour, we should expect a higher casualty rate per employee than per passenger, except in so far as the difference in the nature of the two sorts of returns, as mentioned above, affects the comparison. But the number of employees varies with the number of passengers to be served, hence the inclusion of casualties to employees and, for a similar reason, to "others," in a comprehensive "per passenger" ratio is not indefensible.

REVIEW

1. When is an "injury," resulting in death, termed an accident in the statistical usage of the Public Service Commission? Would this criterion be satisfactory for universal use? Why?
2. What composite units does the author name? Does his contention concerning the definition of these agree with the *Text's?*
3. What are the significant coefficients for Public Utility Statistics of Accidents? In what way is the writer's discussion of this point related to Bowley's treatment of "relativity" in statistical units?

Industrial Accident Rates [1]

The purpose of accident studies is the very practical one of finding out where and why accidents occur and how they may be prevented. The first stage in every such

[1] Adapted with permission from Chaney, Lucian and Hanna, Hugh S., "The Safety Movement in the Iron and Steel Industry, 1907 to 1917," *Bulletin of the United States Bureau of Labor Statistics;* Whole Number 234, pp. 52–66, June, 1918.

study is necessarily the counting and analysis of the accidents reported. In attempting this, two serious difficulties present themselves: First, the lack of a uniform definition of what is to be regarded as an "accident"; and, second, a confusion as to the proper derivation and use of accident rates. Failure to grasp the importance of those two points has been responsible for much loose thinking and many false conclusions, and also has been responsible for the present unsatisfactory character of accident statistics in this country.

Definition of "Accident"

First, then, what is to be regarded as an industrial accident for the purposes of statistical study? No definition has as yet been universally accepted. Some establishments and States attempt to take account of all injuries, however trivial. Others exclude those of a minor character and take account only of such as cause a loss of a specified amount of time. It is evident that the accident showing of a plant may be completely altered by a change in definition of accident, and that in the absence of a uniform definition all comparisons between the accident data of different plants, industries, or other groups become almost worthless. The precise definition is not so important. The important thing is that the same definition should be everywhere observed.

The most significant step so far taken toward such uniformity in this country is the recent action of the International Association of Industrial Accidents Boards and Commissions in adopting a definition of "tabulatable accidents" — *i.e.* a definition not necessarily to be followed in the original reporting of accidents, but to be used in all statistical tabulations. The definition is substantially the same as the one long used by the Bureau of Labor Statistics in its accident investigations and employed in the present report:

"*Tabulatable accidents, diseases, and injuries.*—All accidents, diseases, and injuries arising out of employment and resulting in death, permanent disability, or any loss of time other than the remainder of the day, shift, or turn in which the injury was incurred, shall be classified as 'tabulatable accidents, diseases, and injuries' and a report of all such cases to some State or National authority shall be required."

The States which belong to the International Association of Industrial Accident Boards and Commissions are thus committed to a uniform standard definition of the accidents which are to be tabulated. Some States may at first find it impossible to tabulate all accidents as required by the definition, but the desirability of doing so is apparent, and many have already made a beginning.

The Meaning of Accident Rates.—The second of the two above-mentioned difficulties—the determination and use of accurate accident rates—presents a more serious problem than that involved in definition of accident. Here it is necessary not only to have uniformity, but to decide upon a correct method. In the early attempts of accident statistics, attention was limited to the number of accidents occurring in a given plant or group. But mere numbers, of course, meant nothing unless related to the number of persons exposed to accident. This led to the custom of expressing accident in terms of so many per thousand workers, and constituted an approach to a correct method. To say that a given industry had an accident rate of 100 per thousand workers does convey a definite idea, and can be compared with a rate of, say, 300 per thousand workers in another industry. But the method was extremely crude, because the basic figure "1000 workers" was indefinite and variable. Usually it was derived by rough estimate as to the number of persons employed, such as averaging the number employed at different times of the year or averaging the pay rolls of the year. But no such average could be at all an accurate measure of what was wanted. The number of days worked varies in different plants as do also the daily hours of labor. Two plants may have the same yearly accident rate, say, 200 per "1000 workers," estimated on the above basis, but if one worked only 8 hours a day for 250 days and the other worked 12 hours a day for 365 days, it is clear that the real accident hazard is much higher in the former plant, inasmuch as the same number of accidents per 1000 workers occurred during a much more limited period of time.

Accident Frequency Rates. — From this weakness it became evident that in order to get a rate that would measure real hazard, it is necessary to know not only the number of men employed, but also the time of their employment. The only way to obtain this is to ascertain the actual number of hours worked by all employees for the year. This gives the number of man-hours, *i.e.* the theoretical number of men required to produce the output of the plant in one hour, or what is the same thing, the theoretical number of hours required by one man to turn out the same product. Man-hours so derived constitute the correct basis upon which to calculate accident rates. But the term is unfamiliar and for practical purposes it is convenient to convert man-hours into full-time workers. The full-time worker, as defined by the joint committee of the International Congress on Social Insurance and the International Institute of Statistics, is one who works 10 hours per day for 300 days per annum, making a total of 3000 hours per annum.

The full-time worker, or 300-day worker, so defined, may

seem at first thought to be a mere statistical abstraction. It is true that the full-time worker, like the average man, is a unit of measure, not a living, breathing man, but for the purpose of accident statistics a standardized workman to serve as a unit of measure is absolutely essential. Furthermore, the statistical full-time workman who is assumed to work 10 hours a day for 300 days in the year conforms very closely in most industries to the actual workman who enjoys good health and works every day the establishment is running.

Accident statistics, to be comparable, must be stated in terms of a common unit of measure. The 300-day worker is merely a unit of measure of the quantity of labor, just as the yard is the unit of measure for length. The number of 300-day or full-time workers is obtained by dividing the number of man-hours actually worked in an establishment by 3000, the number of hours per annum assumed to be worked by the 300-day worker.

In those establishments which keep accurate records of the hours worked by each employee every day, the man-hours worked by the establishment can easily be obtained from the records and hence the number of full-time or 300-day workers can easily be computed. Few small establishments, however, keep any such accurate records of time worked. For the majority of small plants it is necessary to compute the number of man-hours worked and the full-time (300-day) workers. The method suggested by the conference called by Commissioner Meeker, which met in Chicago October 12 and 13, 1914, was as follows: "If this exact information is not available in this form in the records, then an approximation should be computed by taking the number of men at work (or enrolled) on a certain day of each month in the year and the average of these numbers multiplied by the number of hours worked by the establishment for the year would be the number of man-hours measuring the exposure to risk for the year."

By the method outlined, true rates are obtained as regards the risk of accident occurrence or frequency. These rates may be called accident frequency rates. Thus if the accident frequency rate, so derived, for the steel industry is 114 per 1000 full-time workers, and is 118 for the machine building industry, it is correct to conclude that accidents are less frequent in the steel industry than in machine building, in the proportion of 114 to 118. All differences in the hours of labor, number of days worked, etc., in the two industries have been duly taken into account. Again, if a given plant shows an accident frequency rate of 100 one year, and 90 the next, it is a correct conclusion that accidents have decreased 10 per cent in frequency.

Accident Severity Rates. — Frequency rates of this character were computed and used in the report on accidents in the iron and steel industry, issued by the Bureau of Labor Statistics in 1913. In all the establishments covered the number of man-hours worked per year was obtained and the working force then reduced to so many full-time or 300-day workers.

The method was found practicable and, within limits, highly useful. But it had one serious weakness, namely, that frequency rates, as the name indicates, measure the frequency of accidents, but take no account of the severity of the resulting injuries, and experience has shown that the two things do not necessarily move in the same direction. The frequency rates may be the same in two plants in the same industry, and the hazards may be entirely different because one plant has very few severe accidents, while the other has a large proportion of serious accidents. To put all industries and all plants on a common basis a system of computing accident rates must be devised which will take into account the difference in economic significance between the accident which bruises the workman's thumb and the accident which breaks his back.

In other words, what is needed is some method of weighting injuries according to their severity. Several methods suggest themselves as possible — compensation paid, wage loss, or time loss. A compensation system necessarily weights the importance of accidents in fixing a scale of benefits which aims to apportion the payment to the hurt. But compensation payments do not offer the universal measure desired because the benefits differ from State to State and are also subject to change within the same State.

Wage loss due to injury offers perhaps a better measure of severity, but this, too, suffers under the handicap that wages differ from place to place and from time to time. Time loss as a measure does not suffer from these objections. An accident that causes 6 days' disability is precisely twice as serious as one causing only three days' disability, and this relation is always and everywhere the same.

The days lost because of injury may thus be taken as the most satisfactory measure of the true hazards of industry — of the burden imposed upon the worker and the community because of industrial accidents. The only difficulty in its practical application is that in case of death and permanent injuries the time lost must be estimated. For temporary disabilities, from which recovery is complete, the time losses are matters of record — 2 days, 10 days, 6 weeks, as the case may be. But, if the accident results in death, the time loss is not so clearly measurable. It exists, however, and may be estimated as the number of working days by which the worker's life was curtailed. Similar estimates are possible in case of permanent injuries, such as loss of hand or foot.

After a study of the available information a table of time losses for injuries resulting in death, permanent total disability, and permanent partial disability was determined upon and applied in this report. The procedure followed was as follows:

Fatalities. — In case of an injury causing death the time loss to the family and society is the expectancy of productive working life of the deceased workman. It is not possible to learn the age of all workmen killed in industrial accidents; but from estimates made by the Wisconsin Industrial Commission, from statistics obtained by several compensation commissions, and from the investigations of the Bureau of Labor Statistics, it seems reasonable to estimate that the average age of victims of fatal accidents is approximately 30 years. According to the American life tables, the life expectancy at age 30 is 35 years. This is for the population as a whole. Workingmen exposed to all the hazards of illness and accident in industry have a shorter expectancy of life than the average for the whole population. The expected productive life of workers is even shorter than their life expectancy. Exact data are lacking, but in the light of all obtainable information it seems fair to estimate the working time lost on the average by relatives and the community for each workman killed by accident as 30 years, or 9000 working days, counting 300 working days to the year. This is admittedly an estimate. A mathematically accurate measure is obviously impossible. It is also unimportant. The main thing is to get the best possible approximation and to apply it to existing accident statistics for the purpose of comparing accident records plant by plant, industry by industry, and year by year.

Permanent Total Disabilities. — If the loss of working time to families and to the community were the sole thing to be shown in accident statistics, the same time loss should be fixed for permanent total disabilities as for fatalities. Permanent total disability is, however, a greater burden to relatives and the community than death. In recognition of this obvious fact the time loss for permanent total disability has been fixed at 35 years or 10,500 working days. The relative importance or burdensomeness of permanent total disabilities as compared with fatalities is thus established rather arbitrarily. After further experience it may be advisable to change the relative weights. The system of weighting used does recognize, however, the undeniable fact that complete permanent incapacity of a worker is a greater burden than his death; and some recognition, even if unscientific, is better than ignoring the obvious facts.

Permanent Partial Disabilities. — A proper weighting for permanent partial disabilities in terms of days lost is even more difficult than for death and permanent total disabilities. An examination of the various compensation acts in existence, however, gives a clue worth following in the quest for some method of estimating the severity of permanent partial disabilities in terms of days lost. First, it appears that all compensation acts agree in fixing the loss of an arm as the most serious injury less than total disability. Most acts, however, seem illiberal in the amount of compensation granted for this injury. The New York act is one of the most liberal. It grants for loss of arm compensation for 312 weeks, which is equivalent to 1872 working days. Inasmuch as the New York scale is based on two-thirds of wages it may be assumed that the entire economic burden was recognized to be one-half greater than the benefit actually allowed. The loss of an arm would thus be equivalent to an economic loss of 468 weeks, or 2808 days. This in turn is equivalent to about 31 per cent of the allowance fixed above for death (9000 days) and 27 per cent of the time lost for permanent total disability (10,500 days). This seemed a reasonable valuation of the arm in relation to permanent total disability and death, and was thus adopted for the scale to be used by the bureau.

Having thus fixed a time value for the arm, it remained to value the other permanent partial disabilities. There is a striking similarity among the various acts in the relation of compensation benefits granted for loss of an arm to those granted for the lesser disabilities. The degree of this uniformity is indicated by the table on p. 174.

Because of the substantial uniformity between the States the scale of awards of almost any State would have given approximately the same relative importance to minor dismemberments compared to loss of arm. The New York scale was adopted as being one of the latest developed, and also because its system of classification of injuries was one readily adaptable to the form in which a large part of the data secured by the bureau was given.

As a result of the above procedure permanently disabling injuries, as well as death itself, were assigned values, expressed in terms of a common denominator—namely, work-days lost. These values, to repeat, are necessarily arbitrary, but the fact that they are not, and cannot be, absolutely accurate, in no way diminishes their usefulness for the purpose in view.

The following table brings together the time losses for death and the more common forms of permanent disabilities as finally adopted for the bureau's scale. Columns of percentages based on this scale of time losses are also given, showing, first, the relative importance of the lesser injuries

TABLE I.—COMPARATIVE TIME ALLOWANCES FOR SPECIFIED DISABILITIES UNDER THE LAWS OF VARIOUS STATES

Weeks for which compensation is payable.

STATE	Loss of											
	Arm	Hand	Leg	Foot	Eye	Thumb	One joint of thumb	First finger	Second finger	Third finger	Fourth finger	Great toe
Connecticut[1]	208	156	182	130	104	38	19	38	30	25	20	38
Illinois[2]	200	150	175	125	100	60	30	35	30	20	15	30
Indiana[1]	200	150	175	125	100	60	15	30	30	30	30	30
Iowa[1]	200	150	175	125	100	40	20	30	25	20	15	25
Kentucky[1]	200	150	200	125	100	60	30	45	30	20	15	30
Maine[3]	150	125	150	125	100	50	25	30	25	18	15	25
Maryland[1]	200	150	175	150	100	50	25	30	25	20	15	25
Massachusetts[2]	50	50	50	50	50	12	12	12	12	12	12	12
Michigan	200	150	175	125	100	60	30	35	30	20	15	30
Minnesota[1]	200	150	175	125	100	60	30	35	30	20	15	30
Montana[1]	200	150	180	125	100	30	20	20	15	12	9	15
Nevada[2]	217	173	195	152	108	65	$32\frac{1}{2}$	39	30	22	17	30
New Jersey[2]	200	150	175	125	100	60	30	35	30	20	15	30
New York[1]	312	244	288	205	128	60	30	46	30	25	15	38
Ohio[2]	200	150	175	125	100	60	30	35	30	20	15	30
Oklahoma[1]	250	200	175	150	100	60	30	35	30	20	15	30
Oregon[4]	416	329	381	277	173	104	52	69	39	35	26	43
Pennsylvania[1]	215	175	215	150	125							
Vermont[1]	170	140	170	120	100	40	20	25	20	15	10	20
Wisconsin[1]	240	160	160	120	120	40	20	20	15	8	10	20

[1] Payments under this schedule are exclusive or in lieu of all other payments.
[2] Payments under this schedule are in addition to payments on account of temporary total disability.
[3] Payments cover total disability. Partial disability may be compensated at end of periods given for not over 300 weeks in all.
[4] The periods named in this line are to be reduced by any time for which payments on account of temporary total disability have been made.

as compared with the loss of an arm, and, second, the relative importance of time losses from death and from the lesser injuries as compared with the time loss from permanent total disability. Other forms or combinations of disabilities than those shown in this list, such as minor injuries to the eye, may be assigned intermediate values. . . .

TABLE II. — TIME LOSSES FIXED FOR DEATH AND PERMANENT DISABILITIES

	Time Losses in Days	Per Cent of Loss of Arm	Per Cent of Permanent Total Disability
Death	9,000		85.7
Permanent total disability	10,500		100.0
Loss of members:			
Arm	2,808	100.0	26.7
Hand	2,196	78.2	20.9
Leg	2,592	92.3	24.7
Foot	1,845	65.7	17.6
Eye	1,152	41.0	11.0
Thumb	540	19.2	5.1
One joint of thumb	270	9.6	2.6
First finger	414	14.7	3.9
Second finger	270	9.6	2.6
Third finger	225	8.0	2.1
Fourth finger	135	4.8	1.3
Great toe	342	12.2	3.3
One joint of great toe	171	6.1	1.6

This schedule supplies a series of constants by which death and permanent injuries may be weighted in terms of a common unit — time lost in days — which is also the same unit as that used for measuring temporary disabilities. Multiplying the number of deaths and permanent disabilities by the time loss determined for each and adding the products to the days lost through temporary disabilities, a figure is obtained which represents the total days lost from injuries. Dividing this number, representing total days lost, by the number of full-time workers gives as a quotient the average number of days lost per full-time worker. This last figure may be called the accident severity rate, since it shows the burdensomeness or seriousness of the accidents analyzed.

The whole process of working out the accident severity rate may be illustrated as follows: Plant A operated 4,200,000 man-hours in 1915, requiring 1400 full-time (300-day, 10-hour-per-day) workers. During the year 324 accidents occurred, resulting in 1 death and the loss of the following members: 2 arms, 1 foot, 5 thumbs, 25 first fingers, while the 290 temporary disabilities showed a time loss of 2790 days. Applying the time losses in the above table to these data, the following results are obtained:

TABLE III.—TIME LOSSES IN ONE PLANT

	Time Loss (in Days)	
	Per case	Total
1 death	9,000	9,000
2 arms	2,808	5,616
1 foot	1,845	1,845
5 thumbs	540	2,700
25 first fingers	414	10,350
290 temporary disabilities		2,790
Total	....	32,301

The total number of days lost, 32,301, divided by the number of full-time workers, 1400, gives an average of 23 days per full-time worker. This is what is here called the accident severity rate, expressed in terms of days. The accident frequency rate for the same group per 1000 full-time 300-day workers would be $324 \div \frac{1400}{1000} = 231$.

ILLUSTRATIONS OF THE USE OF SEVERITY RATES

The preceding paragraphs have explained the meaning of accident severity rates and the method by which they are obtained. The significance of such rates in their practical application is indicated in the two following illustrations:

In the table below comparison is made of the accident experience for a year of the iron and steel industry, as represented by a large plant, and of the machine-building industry, as represented by a group of plants. Frequency rates and severity rates are shown in parallel columns.

TABLE IV. — ACCIDENT RATES IN STEEL MANUFACTURE AND IN MACHINE BUILDING

Industry	Number of 300-Day Workers	Accident Frequency Rates (per 1000 300-Day Workers)				Accident Severity Rates (Days Lost per 300-Day Worker)			
		Death	Permanent disability	Temporary disability	Total	Death	Permanent disability	Temporary disability	Total
Iron and steel (1913)	7,562	1.9	4.6	108.0	114.5	16.6	2.2	2.4	21.2
Machine-building (1912)	115,703	.3	3.6	114.1	118.0	2.9	1.6	1.1	5.6

Examination of the columns giving total frequency rates and total severity rates shows that, on the basis of frequency, the machine-building plants were more hazardous than the steel plant — the respective rates being

118 as against 114.5 per 1000 full-time workers. On the basis of severity, however, the steel plant was almost four times as hazardous as machine building — the days lost per full-time worker being 21.2 and 5.6, respectively. It is clear that as between these diametrically opposite showings of the relative hazards of the two industries, the severity rates offer a decidedly more accurate measure of true hazard. In machine building there is opportunity for many minor injuries, but the danger of serious injury is much less than in the steel industry. The severity rate brings out this fact.

The second illustration shows how, over a period of years, within the same establishment, accident severity rates may run counter to accident frequency rates. The next table gives data of this character. It shows the accident experience of a large steel plant over a period of four years. The plant is one in which most serious attention has been devoted to the prevention of accidents.

TABLE V. — ACCIDENT EXPERIENCE OF A LARGE STEEL PLANT; 1910 TO 1913

YEAR	NUMBER OF 300-DAY WORKERS	ACCIDENT FREQUENCY RATES (PER 1000 300-DAY WORKERS)				ACCIDENT SEVERITY RATES (DAYS LOST PER 300-DAY WORKER)			
		Death	Permanent disability	Temporary disability	Total	Death	Permanent disability	Temporary disability	Total
1910	7642	1.7	4.3	127.5	133.5	15.3	2.4	2.2	19.9
1911	5774	1.6	3.6	106.6	111.8	14.1	2.1	2.4	18.6
1912	7396	.7	6.5	146.3	153.5	6.0	5.5	2.8	14.3
1913	7562	1.9	4.6	108.0	114.5	16.7	2.2	2.4	21.3

Limiting attention to the columns showing total rates, it will be noted that in 1910 the frequency rate was 133.5 per 1000 300-day workers and the severity rate was 19.9 days lost per 300-day worker. The next year, 1911, shows a decrease in both frequency and severity. In 1912, however, there was a marked increase in frequency — from 111.8 to 153.5 — but the severity rate dropped from 18.6 to 14.3. In other words, accidents had considerably increased in frequency, but they were less serious in their total results. In 1913 this experience was reversed. A marked reduction occurred in accident frequency — from 153.5 to 114.5 — while the severity rate jumped from 14.3 to 21.3. In other words, the year 1913, instead of being a "good" year, as it might be assumed to be under the system of frequency rates, was the worst of the four years covered by the table.

These illustrations bring up certain points which it seems desirable to emphasize. The first concerns the use of terms. Severity rates derived in the manner explained are expressed for convenience in terms of work days lost. For instance, the steel plant referred to above is represented as having a severity rate in 1913 of 21.3 days lost per 300-day worker. The term "days lost" as thus used is to some extent a statistical abstraction, but it is close enough to concrete fact to permit of its use in its ordinary sense without any considerable degree of error, provided that the weighting scale employed is a reasonable one. In any case, however, the real significance of severity rates is in their use, not as positive amounts but as relative amounts as indicating the relation between groups. Thus, to recur to the example of the steel plant mentioned, the important fact is that the severity rate for 1913 shows an increase over that for 1912 in the relation of 21.3 to 14.3.

This leads to a second point which cannot be too much emphasized: The fact that inasmuch as the real significance of severity rates is in the measurement of relative hazards, the character of the weighting scale used becomes comparatively unimportant. Thus, by changing the weights in the scale offered above, the resulting severity rates may be considerably altered in their positive amounts, but unless the changes are of a very radical character the relations between the rates for different groups will remain substantially the same. In other words, it is desirable to have the scale used as accurate as possible, but the fact that a completely accurate scale cannot be devised does not impair the value of accident severity rating.

Another fact deserving emphasis is that severity rates have a very important advantage over frequency rates, in that the effect of errors in reporting is minimized. Accident reports are probably never absolutely complete, and, as a rule, the completeness of reporting is in direct proportion to the seriousness of injury. The more serious the injury the greater the likelihood of its being reported. Frequently the reporting of minor injuries is extremely incomplete. Inasmuch as the accuracy of frequency rates depends upon the completeness of accident reports, and as all accidents have the same weight, a failure to report any considerable number of minor accidents renders the rates obtained of very little value. Such is not the case with severity rates. Here the disabilities are weighted according to their importance, and a large group of minor disabilities has comparatively little effect upon the derived severity rate. Thus, from the material available concerning the iron and steel industry, it is estimated that the total exclusion of all disabilities of less than two weeks will rarely diminish the total severity rate for that industry as much as 1 per cent, whereas such an exclusion would diminish frequency rates as much as 60 per cent. In the machine-building industry, according to data collected by the bureau for that industry, the corresponding percentages are 7 and 70.

Growing Recognition of the Importance of Severity Rating. — It is safe to say that all who have been concerned with accident studies and accident-prevention work have felt the need of some system of severity rating, such as that developed in the present chapter. The International Association of Industrial Accident Boards and Commissions has recognized the importance of the subject and through its committee on statistics has the matter now under consideration. The committee has unanimously approved the principle of severity rating. The discussion now concerns simply the scheme of rating to be adopted. The one worked out and applied in the present report is believed to meet the necessary tests of a simple, workable system. It has already been approved and adopted by a number of important establishments.

Use of Rates in the Study of Accident Causes. — Frequency and severity rates, as above described, may be applied to the measurement of accident causes. . . . Inasmuch as the computation of accident rates according to causes is somewhat novel, a brief preliminary description of the method used is desirable.

For any plant, department, occupation, or other industrial group for which the amount of employment and the number of accidents are known, an accident rate may be computed. This total rate may then be apportioned among various causes responsible for the accidents. For example, in a group of blast furnaces, with a total frequency rate of 200 cases per 1000 full-time workers, it was found on analysis that 58 of each 200 cases were due to molten metal, 27 to handling tools and objects, leaving 115 as due to miscellaneous causes. The frequency rate of molten metal as a cause of accident in these blast furnaces was, therefore, 58 per 1000 workers; of handling tools, 27 per 1000 workers, etc.

The value of such rates to the safety man is clearly evident. They indicate, in the example given, that molten metal was the most important single cause of accident in blast furnaces, and the one to which especial attention must be directed.

In the case just cited, the department was taken as the unit, the rates being based on the total employment for the department. If a smaller unit, such as the occupation, be used as a basis, the rates would be based on the amount of employment in the individual occupation. In the case of the above group of blast furnaces it was possible to isolate certain important occupations, to draw accident rates for each, and to apportion such rates among the different causes. Thus it was found that while the frequency rate for the blast-furnace department as a whole was 200 per 1000 workers, the frequency rate for the "cast-house men" was 380 per 1000 workers employed in that occupation. Analysis of causes of accidents showed this total of 380 to be made up of a rate of 201 cases from molten metal, 43 from falling objects, and 136 from "miscellaneous causes."

These occupational cause rates are even more valuable to the safety man than are the preceding departmental cause rates, as they indicate still more precisely the points of greatest hazard. Unfortunately it is not often possible to use the occupation as a unit as plants rarely keep records of employment in such detail, and even if this is done the number of employees in the occupation is often so small as to be inconclusive.

These cause rates, whether based on the department, the occupation, or any other group, are true accident rates, analogous to the death-rates by disease as used in mortality studies. In such studies it is customary to divide the general death-rate for a community into specific rates for the various diseases causing death. Thus a general death-rate of 20 per

1000 for a given city may be made up of the following specific rates: tuberculosis 5, typhoid fever 2, other causes 13. These rates, it may be noted, measure the real prevalence of the several diseases in a way that percentages cannot do. Thus in the year noted, deaths from tuberculosis constituted 25 per cent of all deaths (5 out of 20). Suppose that in the following year a typhoid epidemic increased the typhoid rate from 2 to 7 and thus caused the general rate to jump from 20 to 25, the tuberculosis death-rate of 5 per 1000 would remain as before, but expressed in percentages tuberculosis would have decreased from 25 per cent (5 out of 20) to 20 per cent (5 out of 25) as a cause of death. The percentage change would suggest a great decrease in the tuberculosis hazard, which, however, as the rate accurately indicates (5 per 1000), remained absolutely stationary. The attempt to study causes of death by means of percentage figures is thus liable to be entirely misleading. Rates, on the other hand, offer an absolutely reliable measure. This is equally true, and for the same reasons, in the study of accident causes.

The above illustrations of the use of cause rates were limited, for the sake of simplicity, to frequency rates. Severity rates can, of course, be applied in precisely the same way and with even more valuable results, inasmuch as severity rates, as pointed out above, are a truer measure of accident hazard than are frequency rates.

Use of Rates in the Study of Nature of Injury, Labor Recruiting, and Other Factors. — Frequency and severity rates may also be applied to the study of the nature of injury in precisely the same way as they may be applied, as described above, to the analysis of accident causes. Thus, in a group of blast furnaces, with a total frequency rate of 191 cases per 1000 full-time workers, it was found on analysis that 89 out of each 191 cases resulted in bruises and lacerations, 45 cases in burns, 10 cases in fractures, and 47 cases in various other injuries. This being so, it is quite correct to say that bruises and lacerations in these blast furnaces had a frequency rate of 89 cases per 1000 workers, burns a frequency rate of 17 cases, and so on. These are true rates, with the same superiority to percentages as a measure of the frequency and severity of injuries of various kinds as was noted to be true in the case of accident causes.

Moreover, outside the accident field proper, there are many collateral subjects to which the rate method may be very profitably applied. An important instance of this is the employment of new men. By relating the number of 300-day workers to the number of new men hired during a given time, a rate is obtained which may be referred to as the "labor recruiting rate." There is an interesting and important connection between this "labor recruiting" rate and the accident rate. Usually, the taking on and use of new men has a marked tendency to increase the accident occurrence of a plant.

In a similar manner, rates based on the amount of employment may be derived for production, labor costs, sickness, and many other subjects.

REVIEW

1. Is the author's statement of the purpose of conducting studies of accidents always true? Suggest others.

2. What answer would you give to the writer's question, "What is to be regarded as an industrial accident for the purpose of statistical study?" Can a single definition be given? What relation has the definition to the purpose? Illustrate.

3. What are tabulatable accidents, diseases, injuries? What purpose is kept in mind in deciding this question?

4. What denominators have been chosen in expressing the coefficient "industrial accident rate"? What are their respective

merits? What is meant by a "full-time worker"? How is the unit calculated? Is this a composite unit?

5. What is the method adopted for estimating the "man-hours" worked and the number of "full-time workers"?

6. In summary, explain the expression "accident frequency rate."

7. Explain the expression "accident severity rate."

8. What are the available statistical tests of severity? Are they all equally good? Do they differ for different purposes? Do the interests of the injured, the employee, and the public coincide in establishing such tests?

9. How is the "lost-time" test applied in cases of fatalities, permanent total disabilities, permanent partial disabilities? Does this method appear to you to be scientific? Why? Of what statistical value in this connection is the similarity of the time allowance disabilities in the various States?

10. Calculate the accident severity rate for the following experience, using the schedule of time losses given on page 175.

Man-hours operated per year	5,360,000
Full-time workers	1,800

Accidents — one year.
3 deaths.
1 loss of arm.
1 loss of leg.
1 loss of eye.
60 loss of first joint of thumb.
300 temporary disabilities, resulting in loss of 2670 days.

11. What condition may explain differences in the accident frequency rate between establishments, plants, or industries? What different conditions, if any, explain different accident severity rates?

12. Severity rates are important "not as positive amounts but as relative amounts." Explain. What is the purpose of severity rates in the mind of the writer in making this statement? Might the statement be untrue for other purposes? Illustrate.

13. What relation has the error in reporting accidents to frequency rates, to severity rates? What sorts of error has the author in mind? Might other sorts affect the problem differently?

14. Can you think of any occasions when accident frequency rates would be of greater significance than severity rates?

15. Does the following statement demonstrate the superiority of severity over frequency rates? "Thus, from material available it is estimated that the total exclusion of all disabilities of less than two weeks will rarely diminish the total severity rates for that industry as much as 1 per cent, whereas such an exclusion would diminish frequency rates as much as 60 per cent."

16. Contrast the rate and percentage methods of stating causes of deaths. What relation has the rule of the text, "always relate things to the conditions that produce them" to do with this discussion?

17. Write a single paragraph summarizing the above article and showing its relation to the general topic *Statistical Units of Measurements.*

Some Illogical Units in Railway Statistics [1]

One of the most fascinating and important parts of the statistician's work is the development of the best units or bases of statistical judgment. On this side our published railway statistics compare favorably with any, but none seem above criticism, the principle of coherence is so commonly violated. To be true and logical the unit must be one based on a cause-and-effect relationship, that is, it must vary with the phenomena of whose summation it is an index, or must indicate the relation between worker and work done.

To illustrate this in a negative way take the much overworked train-mile. If it is to be our unit of service, simply as an index of utility rendered, it need only have the quality of varying in proportion to utility consumed by us. Or if it is to have a deeper significance, entering the rate question through the door of cost, it must meet the test of varying with costs, — of indicating the relation between tractive

[1] Adapted with permission from Haney, L. H., "Railway Statistics," in *Quarterly Publications of the American Statistical Association*, September, 1910, Vol. 12, pp. 208–211.

power (the worker) and tonnage movement (the work done). What is the result?

As to service one first reflects that the train is so lacking in homogeneity that the miles it makes are sadly lacking in uniform value. By the time one has asked how many cars there were in that train? what kind of cars — gondolas, box, tank, or stock? at what rate of speed did it move? was it going in the direction of prevailing traffic? was it a train of twenty years ago, made up of 30-ton wooden cars and drawn by a little "American-type" locomotive, or one of to-day with 50-ton steel cars and a Mallet locomotive? — by this time one finds that one train-mile is so different from another that he hesitates to accept it as a standard. Anyhow, he reflects, what one wants from the railway is not train-miles but tons (of goods) moved from A to B, — ton-miles, for short. And while, to be sure, there will be on the average some relation between trains and tons moved, it is not necessary or close enough.

This unit, however, is more often used as a cost index. . . . But, passing over the difficulties of defining a train, it may be said that trains consist of one or more locomotives and a number of cars. In this compound aggregate some costs vary with the number and type of locomotives (wages of engine crew and say 30 per cent of fuel), having no connection with the number of cars, or the "train." Others are peculiar to the cars. Finally there is a remainder that belongs to the train as such (balance of fuel, wages of train crew, etc.). Obviously the train-mile will serve as a homogeneous unit of cost only to the extent that the factors peculiar to locomotive expense and car expense are either negligible or capable of being averaged. Locomotive expenses are far from being negligible. Therefore the value of the train-mile unit partly depends upon an assumed average cost of locomotive-miles, having its weakness. As a unit of cost, perhaps the chief difficulty comes in the varying number and kind of locomotives embraced in the train. Then there are the varying "train resistances," depending on speed, number of cars, grades and curves, etc.

The theoretical lack of relation between train-miles and expense of performance is illustrated by the following relative figures:

Year	Train-miles	Average Cost per Train-mile	Operating Expenses	Ton-miles
1897	100	100	100	100
1902	117	127	148	165
1907	146	158	232	242

To the writer it seems that the usefulness of the train-mile unit varies somewhat according as it is applied to the passenger or the freight service, — suggesting, by the way, that the differences between freight and passenger service cannot be removed through this agency. Considerations other than cost play so great a part in passenger operation that, from the last viewpoint, it has small importance; while in the freight service, if sufficient interpretative data concerning locomotive miles, gross and net tons per train, etc., are utilized, it may be of considerable service. From the service viewpoint the situation is reversed, for in the passenger service train-miles seem to approach more nearly a necessary relation to social service than in the freight service.[1] Considered as an independent unit the locomotive-mile is open to similar objections.

[1] As an index of service between particular points passenger train-miles may be of little value, as they would include trains which did not stop at one or both the points, perhaps, etc.

But without further elaboration the conclusion may be drawn that per-mile units have been too largely depended upon in our railway statistics. Phenomena which do not have a reasonably close relation to miles should not be measured in miles. We need more careful analysis of essential relations and variety of units each adapted to the particular case. A similar weakness might be illustrated from our accident statistics, where the *occasion* is made to serve as a *cause* in some columns.

Take the case of locomotive-miles. As a matter of fact locomotive-hours would mean more; for, taking all expenses connected with locomotive operation into consideration, it will be found that cost varies more with time than distance, —interest, certain repairs, fuel, etc. But hours alone cannot measure locomotive performance; there must be some relation with product. What is produced is "draw-bar pull," or tractive power, so that to really judge efficiency of locomotives from either the cost or the service viewpoint a unit of tractive power must be used. Accordingly a recent report of the Committee on Conducting Freight Transportation of the Association of Transportation and Car Accounting Officers recommends the tractive-power-hour for use by the railways. Thus allowance would be made for different tractive powers, delays between terminals, etc.

Perhaps the true meaning of these different units appears most clearly when the railway is imagined as a great organism whose work is performed through a series of concomitant but subordinate activities. Each department has its function and its product, but that product may be the raw material for another department which carries the work a step farther, — perhaps to its consummation in the final transportation product. Thus in this hierarchy various units may be appropriate for various departments according to their contributions. If it is the function of the terminal force to move cars, "cars handled" is the appropriate unit — of cost, at least. Obviously the ton-mile is not a unit applicable to the work of the mechanical department; that department directly furnishes tractive power on the one hand and carrying capacity — cars, trains — on the other. And so on. The ton-mile caps the climax. But when one desires to judge the particular and peculiar efficiency of a subordinate part of the mechanism its peculiar work must determine its unit. Just as in the case of the cost viewpoint some question might be raised as to how far our government should go, so here it would be necessary to ask how intensive a regulation is desired to determine how many subordinate units are necessary.

Not only is the per-mile average overworked, but also the simple arithmetic average is so used as to be a very relic of barbarism. It is hardly necessary to point out its limitations. As to the particular point now involved it fails in not indicating the weight and distribution of the factors averaged. Why, then, not make some practical use of such well-known statistical devices as the weighted average, the mode, and the median? No average wage for all employees is given; a weighted average would be good. The mode would be best for the average trip and haul. Several shortcomings in the most used units of railway statistics might be capable of partial remedy by the adoption of more illuminating averages, if only the returns were made more analytically.

REVIEW

1. In what way is the discussion of units in this selection related to
(1) the purpose for which the units are used?
(2) the distinction between "simple" and "complex" units?
(3) statistical basis for measuring costs; for measuring "service"?

2. What alternative units to "train-mile" are suggested and for what purpose?

CHAPTER IV

ILLUSTRATIONS OF METHODS IN COLLECTING STATISTICAL DATA

Study of Wages — Method [1]

With a view of supplementing the returns presented in the Report on Manufactures of the Twelfth Census, in regard to earnings of employees making a more precise classification of wages, the Census Office in September, 1901, determined to undertake a special investigation. . . .

1. *Scope and Principles of the Investigation.* — Owing to the limitations of time and the lack of established methods of procedure which could be confidently relied upon, it was determined to limit the scope of the special wage inquiry to a few industries, and to confine the treatment of the data recorded, as far as possible, to a single form. As the method adopted by the Twelfth Census for calculating the number of employees sharing in the total reported earnings differs from that adopted in 1890, so that the data obtained for these two years are not strictly comparable, it was determined to extend the inquiry to 1890 as well as 1900. The principles controlling the investigation are, briefly, as follows:

(1) Restriction of the inquiry to a few stable and normal industries.

(2) Collection of actual *rates* of wages.

[1] Adapted with permission from "Employees and Wages," *Twelfth Census of the United States, Taken in the Year 1900*, 1903, Davis R. Dewey, "Report," pp. xiv–xx.

(3) Classification of employees by rates of wages, and as far as possible by occupations.

2. *Wages as Measured by Earnings and by Rates.* — There are two statistical measures used in representing the reward of labor, commonly termed wages: First, earnings or the income received in a given period of time, irrespective of the number of hours or days actually worked; second, rates which express the amount paid for work during a given unit of time, as an hour, a day, a week, etc. Each of these measures is of value to the student of economic conditions. The first is the compensation actually received in a given period of time without regard to unemployment, occasioned by illness, strikes, industrial depression, or other causes; the second is the earning power in a given unit of time. If employment were regular and constant, these two methods might be used interchangeably — rates could be calculated from earnings and earnings from rates. Employment is not regular and constant, however, because of interruptions due to either individual or industrial conditions. Of the two measures, at the present stage of economic conditions, earnings are of the more interest; but to ascertain the earnings of individual employees for any period of time greater than a week is almost impossible. The earnings as given in the Report on Manufactures of the Twelfth Census, are for a mass of workmen whose identity cannot be preserved from week to week or month to month; as has been seen, the number of employees, among whom the total earnings are divided, is an average number, and to that extent the resulting computations are only approximate.

The earnings of even a single week may be misleading, especially where no record of time is kept by the management. The establishment may have shut down for a portion of a day; work in a particular department of a mill may have been slack, although as a whole the establishment was running full time; or there may have been an exceptional amount of illness at one period as compared with another.

The present inquiry, therefore, is concerned primarily with rates, earnings being used only when the data in regard to rates are defective or require further interpretation. Statistics of rates, however, reveal only a part of the picture; the complete situation can be described only when the amount of time worked for at least a year is known, and even this should be supplemented by a knowledge of prices in order to determine the value of the compensation as measured in the commodities purchased. These latter inquiries must be supplementary; there is no way to combine in one inquiry all the elements for a complete presentation of wage statistics.

3. *The Schedule of Questions.* — In order to carry out the purpose of this inquiry the special schedule on the following page was drafted.

4. *Sections of the Country Covered.* — The work of securing the data called for by this schedule was intrusted to special agents who were instructed to visit certain manufacturing establishments in the respective territories to which they were assigned, care being taken to select essentially manufacturing localities. This restriction, together with lack of sufficient time to make a more thorough canvass, explains the absence of returns from the States classed in the census reports as "Western"; but although the report is to that extent deficient, affording no basis for a comparison of wages between that section and other parts of the country, it is believed that the main results of the investigation are not thereby seriously impaired. Fortunately, returns were secured for a few industries for the Pacific States.

5. *Industries Investigated.* — The inquiry was limited to

TWELFTH CENSUS OF THE UNITED STATES

MANUFACTURES

SPECIAL SCHEDULE — RATES OF WAGES

Name of establishment..

Location..

Industry..

Goods manufactured..

Pay roll of..

Occupation	Number of persons	Sex 16 years and over	Sex under 16 years	Rate of Each		Per hour, day, week, month, or year	Hours of labor per day	Hours of labor per week	Piece or day work	If any employee pays helpers, note same and separate
				Dollars	Cents					

34 industries, nearly all of a permanent character, which are not violently affected by seasonal influences. They are:

Agricultural implements.	Glass.
Bakeries.	Iron and steel.
Breweries.	Knitting mills.
Brickyards.	Lumber and planing mills.
Candy.	Paper mills.
Car and railroad shops.	Pianos.
Carpet mills.	Potteries.
Chemicals.	Printing.
Cigars.	Rubber.
Clothing.	Shipyards.
Collars and cuffs.	Shoes.
Cotton mills.	Silk mills.
Distilleries.	Slaughtering.
Dyeing and finishing textiles.	Tanneries.
Flour mills.	Tobacco.
Foundries and metal working.	Wagons and carriages.
Furniture.	Woolen mills.

In grouping the returns by industries, the plan of classification adopted by the division of manufactures of the Twelfth Census, in which product is the determining factor, has in the main been followed here. For the purpose of analyzing wages in specific occupations this is not a logical classification, as there is no inherent relation between products and occupations; some classification, however, is necessary in order to cover the most important branches of industry, and the grouping by manufactured products is chosen as the most serviceable method available. Almost the only change made in this report in the regular census industry names is a slight alteration of the wording to make them more definitely descriptive of the establishments from which pay rolls have been secured. Thus, the census classification is "tobacco, cigars, and cigarettes," but since no cigarette factories are covered in the present investigation the industry is called "cigars." "Breweries" is used instead of "liquors, malt"; "tanneries" instead of "leather, tanned, curried, and finished"; and other similar changes in wording are made. But in all cases establishments are referred to classes corresponding to those shown in the General Census Reports, except where differences in product would thereby be shown in too great detail. Thus, in the Report on Manufactures of the Twelfth Census, brass foundries, iron foundries, machine shops, bicycle factories, sewing-machine factories, typewriter factories, etc., were given separate classes; but for the purpose of securing the statistics of wages it is believed that the returns can be safely simplified by combining all these as "foundries and metal working," thus obtaining numbers of employees engaged in the same occupations sufficiently large to justify extended study of the results.

The classification for industries is made by establishments as a whole. It has not been considered feasible to attempt to subdivide establishments into departments, except in the case of a few textile establishments, where the books are so kept that the dyeing and finishing departments can be separated. This classification of establishments is presented in four general groups made up of the 34 separate industries. No attempt has been made to consolidate the statistics in these four groups, but in the discussion and arrangement of the statistics the similarities within some of these general classes have been helpful. The industries comprised in the four general groups are as follows:

(1) Textile mills, which comprise reports from carpet mills, cotton mills, dyeing and finishing establishments, knitting mills, silk mills, and woolen mills.

(2) Factories engaged principally in woodworking include agricultural implement factories, furniture factories, lumber and planing mills, piano factories, and wagon and carriage factories.

(3) Metal-working establishments comprise car and railroad shops, foundries and metal-working establishments, iron and steel mills, and shipyards.

(4) Miscellaneous industries reported include bakeries, breweries, brickyards, candy factories, chemical factories, cigar factories, clothing factories, collar and cuff factories, distilleries, flour mills, glass factories, paper mills, potteries, printing establishments, rubber factories, shoe factories, slaughtering establishments, tanneries, and tobacco factories.

Certain resemblances in materials or products might serve as a basis for grouping some of the industries in the last class; thus, for instance, "bakeries," "candy" factories, "flour mills," and "slaughtering" establishments, all furnish foodstuffs; but similarity of product is no reason why they should be grouped in wage statistics. It is not to be expected that two establishments exactly alike as regards labor conditions can be found, but it is believed that within the industries as finally determined, interchange of labor can be accomplished to a considerable extent; that is, each industry represents a group of establishments making similar products by related though diversified processes so that the labor employed in one establishment is comparable with that in another.

The three important steps in wage investigation are collection of data, tabulation, and analysis.

1. *Pay Rolls Copied.* — In the collection of data it was decided to rely upon the pay rolls of employers; only in this way is it possible to secure returns from all the constituent elements in a given establishment, for it is manifestly im-

practicable to visit each separate employee to obtain a personal return; and, moreover, it is clear that the pay roll of the employer states in the most precise form available the actual rate of pay of each employee. This method removes all opportunity for either exaggeration or underestimation and also the possibility of substituting a customary wage for the actual one.

2. *Representative Character of Returns.* — An important consideration in the collection of data is the amount of material required to justify the construction of tables on which reliable conclusions can be based. This question of representativeness of returns is fundamental to the proper development of wage statistics. As it is impossible to secure from every employee a return of his actual wage, so it is impossible to secure a transcript of the pay roll of every manufacturing establishment in the United States. Fortunately, the problem is not so difficult of solution as it may appear. In any given locality there is a strong tendency toward uniformity of wages in the same occupation; if, therefore, the occupations are carefully designated, the number of returns for a given occupation need not necessarily be inclusive of all employees engaged in the same kind of work. The more precisely the occupation is described, with regard to sex, age, and gradations of skill, the fewer are the numbers needed. It is impossible, however, at the present stage of the development of wage statistics, to lay down any definite formula as to the exact proportions required. In this investigation the Census Office has endeavored to secure a harmony in the proportions of returns for different occupations, and believes that for most of the occupations tabulated the numbers are sufficiently large to justify the uses to which they are put.

3. *Selection of Establishments.* — Effort was made, both by the Census Office at the outset and by the agents when

actually on the ground, to select establishments which may be regarded in every respect as representative. It was determined to secure returns from establishments having the largest numbers of employees; and to insure the comparability of the statistics no establishment was chosen which had been in existence less than twelve years. Trial lists of addresses were accordingly prepared from the general manufacturing schedules of 1900 on file in the Census Office. In the progress of the work, however, various practical difficulties arose which made it necessary in some instances to procure pay rolls of small establishments, but in every case, these are well-established undertakings and may safely be regarded as representative. The number of pay rolls utilized in the compilation of the tables is 720. Classified according to the number of employees, the establishments from which these pay rolls were secured are grouped as follows:

Number of Employees per Establishment	Number of Establishments
Total	720
Less than 100	260
100 to 499	336
500 to 999	74
1000 and over	50

4. *Difficulties Met by Special Agents.* — It is gratifying to note that there was a general willingness on the part of employers to furnish pay rolls; objection was a rare exception. The difficulties met by the special agents may be summarized as follows:

(1) Destruction of the pay rolls for one of the two periods: This was due either to fire or to the policy of a company to destroy the pay-roll records after a brief term of years.

(2) Inaccessibility: Sometimes the pay rolls were stored away in attics or cellars, requiring time and labor to make them available. Where the character of the organization had changed, the books of the old concern were often in the hands of some one no longer interested in the operation of the new company. If the old institution had become a part of an industrial combination, with head offices at a distance from the particular plant visited, the superintendent was seldom willing to give the information without authorization by an official of the controlling corporation; frequently in such a case a visit to the head office was necessary.

(3) Imperfect records: Many of the pay rolls were so imperfect that they were worthless for the inquiry. In some of them lump sums were included for contract work without any designation of the number of employees working under the contract; in others the earnings of helpers were consolidated with those of the employees whom they helped. Under these conditions separate wages could not be determined. In establishments where piecework prevailed it was often necessary to ascertain, from small time books kept by the foremen of the various departments, the time actually worked by the individual employee — a task demanding patience and care. Only rarely did the pay rolls separately designate children, even when they were employed, and to determine this point special inquiry generally was necessary; at best the information gathered and returned as to the ages of employees is unsatisfactory, and it is probable that the actual number of employees under 16 years of age is larger than that reported. It was not an infrequent experience for the agents to find by subsequent inquiry that some of the employees returned as 16 years of age and over belonged to the younger age class; only in States where local legislation in regard to school attendance is stringently en-

forced is the classification of age of employees likely to be of much service.

5. *Lack of Uniformity in Pay Rolls.* — The pay rolls which were finally secured are not uniform or simple in character. The two principal sources of difficulty are, first, the variety of time units for which rates are returned; and second, the fact that in many establishments no permanent record of time is kept, and for some of the employees earnings only are reported. Rates are reported by the hour, day, week, half month, and even by the month, or year. Where earnings were returned the time worked in some instances was reported, making it possible to determine the rate; in other cases, however, the time was unknown, and rate tabulations could not be made. . . .

6. *Rejections.* — Whenever the wages returned for an employee include anything besides the actual compensation for his own personal and unassisted services they have been rejected, unless such actual compensation can be definitely determined. For example, the wages of a teamster furnishing his own horses are excluded, and so also is the lump sum reported as paid to a workman with one or more helpers, unless the proportion received by each is given.

Again, where it is evident that the wages reported as paid to an employee were received for work which was additional to and outside of his regular duties, the return for that employee has been omitted. Thus, in the case of a Sunday watchman reported as receiving $2 a week and working twelve hours, there can be no doubt that this wage of $2 is for work additional to and outside of his regular duties, and to show a man who earns $2 for twelve hours' work as receiving only that amount for a week would be palpably wrong.

The wages of persons whose services were chiefly clerical

in their nature are omitted, as are those of all salesmen and superintendents.

Where average earnings are reported, instead of exact earnings or actual rates, such averages are excluded.

7. *Wage Groups.* — In classifying the returns into groups, it is desirable to choose a unit of division small enough to bring out the essential facts. If the group has two extensive limits, it may include employees of widely different grades of skill and compensation, making it difficult to discover changes occurring between the two given periods of time. The ideal method would be to arrange a series of gradations so minute that every employee would be assigned to his actual rate; this, however, is impracticable, both on account of the expense and of the difficulty, under the present limitations of statistical art, of grasping the significance of tables so elaborate in detail. Accordingly, the unit adopted for the tables of this report is 50 cents for week rates and 1 cent for hour rates. Never is a difference of more than 50 cents a week, of 1 cent an hour, necessary to change an employee's standing in the wage scale from one group to another, and often a much smaller difference will produce such a change; thus, for example, when the rate is near the upper limit of the wage group, the amount of increase necessary to remove it to the next higher group varies directly with the distance between the actual rate and the upper limit of the group; on the other hand, the nearer such a rate is to the lower limit of the wage group, the smaller the decrease necessary to cause its removal to the group below.

8. *Time Units.* — The units of time finally adopted as the most serviceable for the tabulation of rates are the hour and the week. The day unit has many advantages, but little information is supplied by day rates which is not found also in hour and week rates. From the week rate it is pos-

sible to determine the maximum amount which a workman can earn per week in normal working hours, and from the hour rate it is possible to discover increases in the rate of wages per unit of exertion which are due to the shortening of the hours of labor per week rather than to an actual increase in the weekly rate of pay. Sometimes, also, the change in the weekly rate is due to a difference in a number of hours worked per week, the rate per hour remaining the same. On account of the variety of the returns great care has been taken in reducing them to a common standard for purposes of presentation and comparison.

It may be remarked that there are several causes which may make the change in the wages of the same persons appear different in the tables of rates per week from those shown by the tables of rates per hour. Briefly stated, these causes are as follows:

(1) The change of normal hours in establishments during the decade.

(2) The combination of returns from establishments with different normal working hours for the various occupations, in which the proportions of the returns of the several establishments change from one period to the other.

(3) The difference in scale between the wage groups in the week and those in the hour tabulations, resulting in a slight change in the distribution of the returns through the groups.

9. *Normal and Actual Working Time.* — Normal time is the number of hours regularly worked under full time. Actual time is the number of hours which a particular employee actually works in earning the amount of money paid him for the period in question. Care has been taken to distinguish between this normal working time for a factory, or a department of a factory, and the actual number of hours worked by each individual employee in that factory

or department. In all cases the rates published are based on the normal time. The only use made of the actual time, when reported, is in the computation of rates from earnings or earnings from rates.

10. *Time and Pieceworkers.* — There are two principal methods of payment for labor — payment for length of time worked, and payment for quantity of work done, or piecework. In the preparation of statistics of wage rates, the wages of time workers are usually returned in practically the form desired for purposes of tabulation, since the basis of payment is a certain amount of money for a certain length of time. For pieceworkers, however, the computation of rates is more difficult; their wages are always reported in the form of the amount paid on the given pay day. Unless the exact time worked in earning this pay is reported, no computation of the wage rate is possible; but when the working time required to earn the pay reported is stated, the computation of a time rate is considered justifiable. For while piecework may be described as a system under which an employee sells to his employer a specified quantity of labor, irrespective of the time occupied in the performance of that labor, and time work as a system under which he sells to his employer the labor which he shall perform within a given period, irrespective of what the quantity of that labor may be, yet in each case both the time worked and the quantity of work done are taken into consideration in fixing the rate of pay. A piece rate always implies a time basis, being adjusted with reference to the time required by the average workman for the performance of a given piece of work; conversely, a time rate always implies a piece basis, for the workman under this system must usually perform a certain minimum of work or lose his place. Thus the two systems of payment, although

apparently diverse, are so closely related as to warrant the computation of time rates for pieceworkers when the exact working time of the pieceworker is reported; especially is this true for purposes of comparison.

11. *Necessity for Computation of Rates.* — Each line of a pay-roll schedule shows the rate per hour, day, week, month, or year, in some cases per two weeks, and in one or two instances per quarter hour, for one or more employees doing the same work and receiving the same wage. As the purpose is to present tables showing rates per hour and per week (or when this is impossible, earnings per week), it is necessary, when one is given, to compute the other, and when neither the week nor hour rate is given to compute both from the data that are given. A considerable number of pay rolls show earnings for the period covered by them — *i.e.* a week, two weeks, or a month, as the case may be. This is, of course, the rule when returns are made for pieceworkers. In such cases the rates per hour and week can be derived by computation only when the exact number of hours worked is stated or the actual number of days of known length is given. The time worked to earn the amount given is never estimated, no attempt being made to derive rates from earnings unless the number of hours worked to earn the amount stated is definitely known for the individual employee.

12. *Rules for Computation of Rates.* — The following are the general rules according to which the computation of rates is made:

(1) When the rate given is per hour, the week rate is obtained by multiplying the hour rate by the number of hours regularly worked in a week by the employee.

(2) When the rate given is per day, the hour rate is obtained by dividing the day rate by the number of hours regularly

worked in a day, and the week rate is then obtained as in (1). (For exception see section 14, below.)

(3) When the rate given is per week, the hour rate is obtained by dividing the week rate by the number of hours regularly worked in a week.

(4) When the rate given is bi-weekly, a weekly rate is obtained by dividing the bi-weekly rate by 2, and the resulting rate per week is then treated as in (3).

(5) When the rate given is per month, unless for an employee regularly working every day, including Sunday, a day rate is obtained by dividing the monthly rate by 26, and the day rate thus obtained is treated as in (2). In cases where a monthly rate is given for an employee regularly working every day in the week, including Sunday, the rate per day is the result of dividing the rate per month by 30 instead of by 26.

(6) When the rate given is per year, it is first reduced to a monthly rate by dividing by 12, and the monthly rate thus obtained is treated as in (5).

13. *Exception for Iron and Steel Industry.* — The prevalence of turn or tour duty in the iron and steel industry makes necessary some slight exceptions to the general rules adopted for the computation of wages in other industries. In this industry a turn, tour, trick, or shift is 12 hours long in many establishments, one crew working from noon till midnight and the other from midnight till noon. The night crew in a number of plants works only 5 days a week, and as those who work at night one week work during the day the following week, an employee puts in only 11 days in two weeks. This constant and regular variation in the normal working hours per week for many establishments makes it advisable to compute rates for the operative in this industry on the basis of 2 weeks instead of 1, and this

has been done. For such employees as work in turns, 6 days in one week and 5 the next, a day rate is obtained and multiplied by 11, while for those who work 6 days in each week, the day rate is multiplied by 12. Otherwise the rates are computed according to the general rules already given.

14. *Exception for Half Holiday without Loss of Pay.* — Pay rolls were submitted by some establishments which paid their employees for 6 full days although the plants closed early on Saturday — at noon in some cases. The rates for this class of establishments are somewhat differently computed; if an hour or day rate is returned, the week rate is obtained by multiplying the rate given by the number of hours or days, as the case may be, in a week of 6 normal days. The week rate so obtained is then, for a new hour rate, divided by the number of hours normally worked. For example, a machinist may be paid 30 cents an hour for 10 hours a day, 60 hours a week, although the plant where he is employed closes regularly at noon on Saturdays. The number of hours actually worked by this machinist each week will be, then, not 60, but 55. Since he is paid for a full week, he really receives $18 for 55 hours' work, 32.7 cents an hour, although if he worked anything less than full time he would receive compensation at the rate of 30 cents an hour. He stands in the same position, as far as earnings are concerned, as the machinist who is paid 30 cents an hour, but who must work 60 hours a week; both receive $18 a week, but the first gets, in addition to his money wages, a certain amount of time which is his own. This advantage is usually, if not always, made contingent on the operative working full time, but as rates are always computed on the basis of full normal time, that fact is not here material. Other things being equal, the first, working 55 hours a week, enjoys an advantage over the employee

working 60 hours, and to show this advantage the above exception to the ordinary rules of computation is made.

15. *Computation of Earnings.* — The pay rolls showing earnings without giving the actual time worked by the wage earner, although of secondary importance, are deemed too valuable to be disregarded, and the returns of earnings have therefore been presented in separate earnings tables. The only period for which actual earnings can be accurately ascertained is that for which they are reported, namely, the period covered by a single wage payment. In most cases this is a week, but, as in the case of rates, there is some diversity, the period being sometimes a half-month or a month.

For the purposes of this inquiry the week is a more satisfactory period than the month, as well as a more available one. In any large factory there will be a considerable number of men who will be found to have worked full time, whether the period be a week or a month; but of those who may be considered regular employees, more will have been absent some time in a month than in a week, and there will also be more old hands discharged or new ones taken on, or both. Moreover, in a month the number of short-time men will be greater than in a week, and consequently the total number of employees reported will be larger. The aggregate amount of lost time will probably be about the same in one week as in another, apart from any general shutdown in the entire factory, and the period including such a shutdown would not be selected by the special agent. Consequently it is believed that the computation of earnings for a week from the reports for a longer period is justified.

For these reasons the week has been adopted as the basis for the tabulation of earnings, and where the earnings reported are for a longer period they are reduced to the week basis. To the objection that such a reduction should not be made, it is answered that the reduction made in the present investigation is justified by two facts: First, the number of returns to which this objection would apply is very small; and second, the special agents in taking these long-time pay rolls usually omitted the employees who worked only a small part of the pay period. These considerations have no effect on the computation of rates, but if the reduction of earnings for a month to earnings for a week were more frequent it would affect unfavorably the value which the earnings statistics might have. The rules according to which the earnings computations are made are as follows:

(1) When earnings are stated for a two-week period, those for one week are obtained by dividing by 2.

(2) When earnings are stated for a month, they are divided by 26, the number of working days in a month, and the resulting quotient is multiplied by 6. In cases where the wage earners work regularly 7 days a week the divisor used is 30 instead of 26, and the resulting quotient is multiplied by 7 instead of by 6.

(3) When rates are returned with the exact time worked, in addition to the time normally worked, then, after the card is computed for rates, the earnings are obtained by multiplying the rate per hour by the exact number of hours worked in the period covered by the pay roll, and if for a period other than a week they are reduced to a weekly basis.

16. *Computation of Percentages.* — In working percentages computations are carried to two places of decimals, and the second allowed to influence the first, which is the last figure shown. In the case of cumulative percentages the accumulation is first made and the resulting percentage shown to one place of decimals.

REVIEW

1. What distinctions are made between the names which are used to describe the compensation which employees receive? Do these agree with those formulated in the *Text*, Chapter IV? What difficulties are mentioned in securing records of compensation? Do these seem real to you? Why?

2. What bases are used in grouping the industries for tabulation? Do these seem logical to you? Suggest others. What conditions seem to have determined the grouping used?

3. Under the headings "step in wage investigation," "collection of data," what topics are discussed?

4. What difficulties were encountered in the use of pay rolls?

5. What problems are suggested in the contention that the returns must be representative?

6. What things were considered and why in fixing the wage groups for tabulation? In fixing the time units for expression of wage data? Do the considerations noted here seem to you to be of general application, or are they limited to this particular statistical problem?

7. Why were the rates published based on "normal time"?

8. How were the piece rates reduced to a time basis? Is such reduction always possible?

9. What rules were followed in computing "*rate* of compensation"? Why was a *week* chosen as the rate period?

Statistics of the United States Shipping Board [1]

I. *Introduction*

What is said about the statistics of the United States Shipping Board has to do primarily, but not solely, with the Division of Planning and Statistics.

The Division of Planning and Statistics of the Shipping Board, at the time of its organization, was unique among

[1] Adapted from Secrist, Horace, "Statistics of the United States Shipping Board," *Quarterly Publications of the American Statistical Association*, March, 1919, pp. 236–247.

government bureaus. It was created in response to an urgent need for the development of a plan and method in the utilization of American and American controlled foreign tonnage in the prosecution of the war. Foresight and planning were to be and have been the guiding principles in its development. The making of history, the production of finished and comparable statistical reports have constantly been sacrificed to the need for day-to-day statistics of use for planning purposes. Hence, statistical hazards — jumps in the dark, as it were — were taken when there was only the smallest chance of their being justified when viewed from any other angle than the emergency which prompted them. As fast as conditions were standardized the statistics were improved; they became more comprehensive, and more closely followed the canons imposed by approved methods.

II. *The Problems to Be Met by the Division*

At the beginning of 1918, the United States had a small merchant fleet of its own, a nascent emergency fleet, some enemy seized and requisitioned neutral vessels. Both in number and in manner of use, they were inadequate to guarantee a "bridge of ships" either for war or trade purposes. Moreover, to leave them in their accustomed trades would only aggravate the shortage. Control of imports first, and of exports later was imperative. Moreover, Government control of vessels was necessary. This was provided by requisition orders and covered not only vessels building in the United States on American and foreign account, but also vessels trading between the United States and foreign countries. Control both of vessels and of commodities seemed to guarantee against a wasteful use of United States and United States controlled foreign shipping. Administrative action and intelligent planning, however, were necessary in order that economical use might be realized. How was this secured so far as the Division of Planning and Statistics of the Shipping Board is concerned?

Study by the Commodities section of the Division unmistakably revealed the importation into the United States of "unnecessary" goods. Such use of ship tonnage could not be defended in any scheme which made "win the war by intelligently using ships" its chief sanction. This fact was patent, but to measure the amounts in long tons of such unnecessary imports, often quoted in trade statistics in values or in containers, and the equivalent ship tonnage "wasted" through such importation presented real statistical problems. These were the first, and continued to be some of the most difficult, statistical problems of the Division.

By statistical analysis, consultation with the trade, with the Army, the Navy, the Food Administration, the State Department, and other Government agencies, an import program was finally established. In outline, this provided that the War Trade Board should license imports and that the Shipping Board should provide the necessary ship tonnage to move them. In working out this program, trade protests and diplomatic objections had to be met or circumvented. The argument, that to cut imports saved ship tonnage, was true, but its application was neither seen nor welcomed at first by the interests involved. The question was asked — and later, answered — How much tonnage? It was necessary to determine the amount of savings not only to meet trade objection, but likewise to furnish a basis for the assignment of tonnage by the Shipping Board so as to guarantee that the import program in its civilian and army aspect would be met. To answer the shipping side of this question required that the following, among other problems, be studied, and that the results of such study become controlling factors in the daily administrative routine of the Division, and of the other war boards with which it coöperated.

(1) The stowage of goods in space and weight.

(2) The conversion into long tons of the values and other units in which imports are often expressed.

(3) The turn-arounds of vessels, or the time spent in completing one round trip.

(4) The unit in which to measure cargo capacity, in the study of vessel utilization.

(5) The relations between the ship tonnages in current use.

(6) The relations between the different types of vessels as carrying units.

(7) The relations of bunkers and stores to total ship tonnage in order to determine the capacity for cargo tonnage.

(8) Use and practicability of combination as contrasted with solid cargoes, and the relation of the distribution of necessary imports thereto.

(9) Suitability of vessels for various services, account being taken, among other things, of size, speed, permanent bunkers, fuel consumption, charter restrictions, etc.

(10) Ballast movements, and underloading by space and weight.

(11) Distribution of ship tonnage by trades and services.

(12) Vessel control by flag, charter, agreement, etc.

(13) Losses through marine risk and enemy action.

(14) Acquisitions to merchant fleets through building, purchase, charter, repair, and salvage.

This list, though far from complete, will serve to illustrate the types of problems with which it was necessary to deal. The statistical material for their measurement had, for the most part, to be created, or secured in a crude state from widely different and frequently conflicting sources. A review of this in terms of the problems named, may be interesting. It is impossible, however, in this short paper, to develop fully any of these topics, and to criticize from a statistical point of view the sources of material and the uses to which they are put. Little more can be done than to list them.

III. *Sources of Material*

(1) "Cargo Reports."

These are reports made by masters of vessels to collectors of customs on vessels, (*a*) entering foreign, (*b*) clearing foreign, (*c*) arriving coastwise. They are made in duplicate, one copy going to the Shipping Board, and one being filed by the collector. They show (using the entering form as an example), for individual vessels, the port of entry, name, type, flag, port of origin, date of clearance, ports of call, with arrival and clearance dates; gross, net, and total deadweight-tonnage; tons of bunkers, water, and stores on leaving; days spent in port of origin; deadweight for cargo; total cargo on board in long tons and cubic feet; total capacity in cubic feet (bale and grain); description of cargo, showing for each of about sixty principal commodities, port of loading, long tons on board, and cubic feet of space employed; amount to be discharged at port of entry in long tons; etc.

These reports are fundamental, and supply source material on stowage, tonnages, source of imports, solid and combination cargoes, turn-arounds, delays in port; ballast movements and vessel utilization, relation of total to cargo deadweight, etc.

(2) Application for License for Bunker Fuel, Port, Sea, and Ship's Stores and Supplies. — "Bunker Form B-1."

This is a report made out in triplicate by the owner, charterer, agent, or master of a vessel, and is presented to the agent of the Bureau of Transportation of the War Trade Board or to the collector of customs. One copy goes to the Shipping Board. Among other things, this report calls for the name, flag, type, speed; registered gross, net, and total deadweight-tonnage of vessels; average daily consumption of fuel in port; owner's and charterer's name, address, and nationality; date of charter party; date of expiration of charter party; trading limits, if on time charter; ports of call on last completed voyage; last port outside United States from which vessel cleared; description of complete voyage which is to be made; etc. This report when sent to the Shipping Board also contains a statement of the amount of fuel and stores actually licensed to be put on board.

This report, likewise, is fundamental in the work of the Division, throwing light, not only on the characteristics of vessels, but also on their control, trading limits, and most distinctive of all, on the relation of coal consumption to the voyage in question. By means of it, the steaming radius of a vessel and the relation of total to cargo deadweight are checked against other sources, or independently determined.

(3) Master's Report on Outward Voyage. "Bunker Form B-3."

This is a report made out by the master of the vessel at the time of completing his voyage and provides for his listing all ports of call with dates of arrival and departure; cargo and bunkers loaded and discharged, and the amount of fuel on board at place of destination. The report, although not received until the voyage is completed, is important in tracing vessel itineraries and periods of turn-around.

(4) "Charter Reports."

The Chartering Committee of the Shipping Board approves charter parties of American and foreign vessels doing foreign business with the United States and of foreign vessels doing coastwise business. A daily report on charters, approved, disapproved, and cancelled, is made to the Division, and gives, in addition to descriptive facts of vessels, the names of owner, chartered owner, operative charterer; form and duration of charter; trading limits, if on time charter, etc. By means of this and other reports, record is kept and studies made of American vessels chartered to foreigners; foreign vessels chartered to the United States Shipping Board or United States citizens; foreign vessels chartered to foreigners under conditions approved by the Chartering Committee; and foreign vessels trading with the United States which are specifically required to return to the United States.

(5) "Allocation Sheets" — Ship Control Committee.

Reports from the Ship Control Committee for the Shipping Board at New York are received daily. These show the allocations daily made, the operative companies, and for trans-Atlantic regions, vessels en route each way; those in home ports; those in foreign ports, and the account upon which each is moving. Somewhat similar, but far less satisfactory, reports are received from the committee on vessels trading with South America, the West Indies, and Caribbean points, and in the Pacific.

(6) Reports from the Division of Operations, United States Shipping Board.

The Division of Planning and Statistics relies on the Division of Operations for a large amount of data on the measurements of vessels, ownership, assignment for operation, and charter relations to the board. These and other data are made available through printed or mimeographed reports, or through daily digests of the correspondence of the Division.

(7) Reports from the United States Shipping Board Emergency Fleet Corporation.

Likewise, the Division receives from the Emergency Fleet Corporation, among other things, daily reports on keels laid, launchings, and deliveries, contract measurements of vessels. Actual measurements are later substituted after vessel trials are made, and the itineraries, loading factors, and general utilization of Emergency Fleet vessels watched in the same way as they are for others.

(8) Reports from the Bureau of Navigation, Department of Commerce.

The monthly and yearly reports on American vessels documented, registered, given signal letters, and otherwise listed by the Bureau of Navigation are exceedingly helpful in developing records of our own merchant marine. Moreover, the bureau's reports on shipbuilding and losses are helpful in distinguishing private from public building, and for purging the Division's files of vessels lost through marine risk and enemy action.

(9) Telegraphic Records of Vessel Movements.

A. Cablegrams from American Consuls.

Daily cablegrams are received by the Division from certain foreign ports, and weekly cablegrams from others, giving name, flag, and principal cargo of vessels arriving from or departing for the United States. This information is significant for purpose of vessel loading and allocation, for determining the degree to which the import program is currently being met, and for providing cargoes at home and at foreign ports.

B. The Naval Communication Service.

The Navy Department, through the Naval Communication Service, secures daily by telegraph or cable, information on arrivals at and departures from American and from a number of foreign ports. This information is distributed in printed form daily, and constitutes, for operative purposes, probably the most significant single source of information on vessel movement available to the Division. The facts given include name of vessel, flag, net tonnage, dates, and places of departure or arrival. Occasionally facts on cargo are also included, but these are far too meager and uncertain to serve as satisfactory data on this topic.

C. Other Cablegrams.

Cablegrams to and from the State Department, War Trade Board, Shipping Board, Division of Planning and Statistics, Division of Operations, and the Ship Control Committee serve currently to correct the files of the Division on the operative status, charter and ownership control of vessels, to indicate the types of problems that are to be solved, and to suggest statistical summaries and reports which are helpful to that end.

These sources of information and the problems upon which they bear have to do primarily with the domestic side of the shipping problem, in so far as it is handled by the Division of Planning and Statistics. There is, however, the international side which should receive attention, both as to source of material and the problems involved.

IV. *The Division and the American Section of the Allied Maritime Transport Council*

As events have turned out, the Division of Planning and Statistics is the primary agent through which American shipping facts are furnished to the American Section of the Allied Maritime Transport Council, and to the allied nations generally as represented in the Secretariat of the Council itself. Early in the summer of this year, it became evident that the American Section of the Allied Council was not currently receiving from the United States the material that it needed to present fully the shipping situation of the United States at the meetings of the Council in London. Mr. Rublee, one of our representatives in London, came to the United States in June of this year to present the case of the American Section and it was not until the time of his visit that the obligation of the Shipping Board to the American Section was fully realized. Domestic affairs, the newness of the work of the Division, the paucity of records, and the insistence of those at home for information all served to keep the outlook of the Division domestic. Following Mr. Rublee's visit, however, Mr. E. F. Gay, director of the Division, sent the writer to London to study the needs of the American Section as they were related to tonnage matters, to provide machinery for meeting them, to determine the ways in which the American and other sections of the Council could serve the Shipping Board, and to establish the necessary connections and the required machinery for securing these services. Later Mr. W. S. Tower of the Commodity Section of the Division was likewise sent to London to study the import and export phases of the problem.

As a result of these visits, and of the more thorough

knowledge of the problems of the American Section and of the Shipping Board, a large part of the activities of the Division has been devoted to a consideration of the shipping problem in its international aspect. Information on the composition of the merchant marines of the allied, enemy, and neutral countries; on movements and cargoes of allied vessels; on losses through marine risks and enemy action; on shipbuilding, repairs, salvages, charter rates, and amounts of chartered tonnage, etc., is furnished this Division by or through the American Section. Information is also supplied by the British Ministry of Shipping, the British Admiralty, and Lloyds. Some of it comes by cable, and some by embassy pouch, but it is all illuminating to the shipping problems of the world and vital in the determination of our part in them.

In supplying information on shipping problems, the Division of Planning and Statistics fully reciprocates. It sends the American Section, and through it the allied countries generally, either by cable or pouch, current data on American shipbuilding, American losses through enemy action and marine risk, repairs to American and foreign vessels, employment of American vessels and foreign vessels controlled by us, inventory facts on American Merchant Marine; required imports in long tons, and the ship tonnage necessary to move them, together with statements in detail of the types, flags, charter relations, and performances of the vessels involved. These reports are by individual vessel, as well as by aggregates, and follow the forms drafted by the representatives of the Allied Council as bases for employment, and loss and gain statements.

So long as the shipping problems of the Allies are adjusted by an international council, the Division can expect to receive from and to furnish to the American and other sections of the Council current information on merchant shipping. The open, frank, give-and-take philosophy which has characterized the relations of the Council and the Shipping Board is illustrative of the unity of purpose with which nations will associate themselves for a common end. As a result of the coöperation, the American Section in London is compiling master files of American vessels (it has full access to the files of the British Ministry of Shipping, Intelligence Branch) and the Division of Planning and Statistics has built up both master and movement files on practically the entire sea-going merchant tonnage of the world. It not only has developed the machinery for efficiently prosecuting the war, but also has collected facts which, if continued, will be of value in promoting trade.

V. *The Division and Other Government Departments*

The coöperation of the Division with other Government Boards should be briefly mentioned. Probably the Departments with which it most fully coöperates are the War Department and the Ship Control Committee of the Board. It furnishes both organizations periodic employment statements covering American and foreign controlled vessels, and special studies of vessels suitable for use in Army service, when judged by standards of physical capacity, charter limitations, etc. A semi-monthly ship balance sheet of tonnage employed and required serves to show not only tonnage distribution, but also the nature of the excesses and deficiencies in tonnage, in trade, and in Army uses. From this statement, the Army knows currently the amount and character of tonnage in trade and is in a position to present its case for transfer of vessels to war use. Similarly, the Ship Control Committee is able to view the trade situation vessel by vessel, and as a whole, and intelligently to allocate tonnage between trades, commodities, and special services.

The Division, too, has closely coöperated with the War Trade Board, in the administration of the Trading-with-the-Enemy Act, and in the collection and preparation of shipping data on foreign countries as a basis for negotiating and administering trade and shipping agreements.

VI. *The Division's Activities Illustrated by Periodic and Special Reports*

The scope of the Division's activities may be further illustrated by listing a few of the many subjects covered in its statistical reports and memoranda.

1. Employment of United States vessels and foreign vessels controlled by the United States by type, form of control, by trade use, and assignment.

2. Private and public charter control of foreign vessels and vessels under agreement with the Shipping Board.

3. Utilization by space and weight factors of vessels arriving and clearing foreign.

4. Merchant marine of the American and the principal foreign countries 1914 to date, showing the losses and gains by causes, trade distribution, and movement.

5. Internment and seizures of enemy vessels in American and foreign ports.

6. Trade of English controlled tonnage between South America and England; between Australasia and all parts of the world, between Africa and northern Europe.

7. American coastwise vessels and commodity movement.

8. Import and export distribution of American and American controlled foreign tonnage.

9. The employment of the merchant fleets of Holland, England, etc.

10. Performance of vessels built for the Emergency Fleet Corporation.

The above reports are illustrative only, and in no way exhaust the topics upon which reports are periodically and occasionally made.

VII. *The Division as a Repository of Shipping Information*

A word should be said of the Division as a repository of shipping information. The significant descriptive facts of the merchant marines of the important countries of the world are in the files of the Division. Moreover, they contain for practically all American vessels, and for foreign vessels controlled by the United States, the itineraries from April 1 to date, adjusted to a graphic scale, distinguishing time in port for ports of entry and clearance, time at sea, cargo carried, and space and weight utilized. Similar facts, but less complete, are available for practically the entire merchant fleet of the world, from June, 1918, to date, whether trading with the United States or not. The cargo, bunker, and master's reports contain basic data for far more complete studies on turn-arounds, loading, ballast movement, port delays, etc., than it has been possible to make during the war. It is the hope of the writer that these data, which have been of distinct service in the control and utilization of our merchant fleet during the war, will be more fully utilized for the development of shipping facts vital to the peaceful prosecution of trade.

VIII. *The Division in Peace Times*

Concerning the peace functions of the Division, a word is necessary.

Changes in source material and in methods will be necessary in order for the Division to retain during peace its distinctive and unique character. These changes must be made in the same thoughtful manner that was used in placing the Division on a war basis. There is room among the present trade and commerce bureaus of the Government for a Division of Planning and Statistics of the Shipping Board, but in order to guarantee against serious overlapping of function, jealousies, conflicts of jurisdiction, and waste of public money, the same readiness to adjust means to ends which has characterized the work of the Division during its year of activity must be adopted by all of the trade bureaus having to do with foreign commerce and shipping, and out of their coöperative endeavor must come a new alignment of function and duties in order to guarantee from each distinctive and unique contributions.

Points to Be Considered in the Use and Form of Questionnaires [1]

Object of an Inquiry. — A problem is half solved when it is clearly stated. Write yourself a memorandum stating what action depends upon having this information; show how the action hinges. Outline your plan for translating the replies into shape for decisive action.

Existing Data. — Before starting anything now, find out what has been done already. This covers, in the first place, your own offices; then, the regular peace time statistical offices of the Government; third, the reliable sources of trade statistics; and fourth, the special investigation by war agencies.

[1] Adapted with permission from *Weekly Statistical News*, Central Bureau of Planning and Statistics, Washington, D. C., No. 9, Nov. 8, 1918, pp. 4–7.

1. *Standard Size.* — So far as possible use $8\frac{1}{2} \times 11$ paper (or multiples of this size, if necessary). This will not only be most convenient to file but also will enable the respondent to expand the report when necessary by adding extra sheets of commercial size typewriting paper. Occasionally it will be desirable to use a small card which can be filed directly in a card catalogue. This device should be used only after very careful consideration of all the limiting factors. The filing equipment to be used must be considered, also, the arrangement in the files and the arrangement on the card adjusted to facilitate filing and finding, etc.

2. *Medium Weight Stock.* — When questionnaires are printed, a medium weight paper should be used. It should be heavy enough to handle easily and to stand well in the files.

3. *Watermark.* — Prefer a paper without a watermark, so that blue prints may be made directly from the original should it become desirable.

4. *Typography.* — Forms should be printed rather than mimeographed, except in emergencies.

5. *Separate Sheets.* — If the questionnaire covers several sheets do not fasten them together in a book, as this makes it difficult, if not impossible, to utilize the typewriter and the carbon paper process of manifolding.

6. *Binding Margin.* — Leave a sufficient margin for binding, preferably at the side, but at the top when wide tabular arrangements are necessary.

7. *Title.* — Each questionnaire should have a distinctive title, which should be as brief as possible, to facilitate reference, etc. It should include the name and address of the office issuing the questionnaire and some indication of its scope. Usually the report should be as of a given date, or covering a specified period.

8. *Sheet Identification.* — Each sheet should carry data adequate to identify it in the event of its becoming separate from its fellows, *e.g.* form number, name of respondent, and date of report.

9. *Pagination.* — In the upper corner opposite the binding side of the sheet, place the page or sheet number. If binding margin is at the top, place page number at the bottom.

10. *Column Designation.* — Where a columnar form is likely to extend beyond one page, designate the columns by letters or figures so that sheets of plain paper may be added by the respondent, using the letters in lieu of printed box headings.

11. *Question Designation.* — So far as practicable number or letter each question and each distinct part of a question so as to abbreviate reference in correspondence. In general, letter the columns and number the rows.

12. *Typewriter Limitations.* — Facilitate the use of the typewriter by adjusting spaces, etc., to meet the limitations of standard typewriters. Horizontal lines should be one-sixth of an inch apart or multiples of that distance.

13. *Abbreviation.* — So far as possible arrange entries which must be repeated so that a brief identification will take the place of a long description in all entries after the first.

14. *Unit.* — Make sure that the unit of every denominate number will be clearly indicated on the return.

15. *Standard Unit.* — Whenever possible specify the unit to be used, so that the returns can be tabulated without conversion.

16. *Common Unit.* — Whenever an entire page, column, or line with several entries is devoted to statistics of a single denomination, show the unit once for all at the beginning of the page, column, or row.

17. *Arrangement in Categorical Entries.* — Let the general precede the specific; the whole, the part; etc.

18. *Position of Instructions.* — If the instructions are not too voluminous, they should appear each at the point where it is applicable.

19. *Arrangement of Instructions.* — Care should be taken to arrange instructions in the order of execution.

20. *Designation of Instructions.* — When it is necessary to separate instructions from their related questions, the instructions should be numbered or lettered to facilitate reference. (N.B. If the questions, etc., are numbered, the instructions should be lettered, and vice versa.)

21. *References to Instructions.* — Insert references to specific instruction in box headings, etc., when it is not practicable to print them in position.

22. *Ambiguity.* — It is not enough that the expressions used reflect the picture in the mind of the author, they should be such that the reader must perforce visualize the same picture.

23. *Terminology.* — So far as possible use terms which are familiar to the respondents. Employ standard terms where standards have been fixed. Define all terms which otherwise might be employed or understood in more than one way.

24. *Tabular Arrangement.* — Frequently a tabular arrangement, combining several questions, not only saves much verbal repetition in the questions, but also makes the logical relation clearer and facilitates the work of answering.

25. *Form of Answer.* — In general, questions should be in the form best adapted to facilitate answers. Give preference to questions which can be answered by "Yes" or "No" or by a number. If answers are to be given by checking or crossing out words explain clearly which prac-

tice is to be followed. Arrange the typography to facilitate the method and stick to the one method throughout the entire form.

26. *Columns.* — If numbers are to be entered which have to be added arrange the questionnaire so that the numbers will fall into columns.

27. *Calculations.* — As a rule, do not ask the respondent to do arithmetic.

28. *Estimates.* — When the obtaining of exact quantities involves great labor, consider whether estimates cannot be used instead. If such is the case, state clearly that an estimate will suffice.

29. *Articulation.* — So far as possible, make the questions such that the answers must corroborate each other.

30. *Letter of Transmittal.* — In practically all cases the questionnaire should be accompanied by a letter covering the general situation; when the data requested are few, the letter may be placed on the upper half of the sheet and the questionnaire below. In such cases do not fail to inclose a duplicate for respondent's file.

31. *Typography of Letter.* — The general appearance of the letter should be such that it will not be confused with advertising matter. The multigraph is to be preferred to the mimeograph for such letters.

32. *Tone of Letter.* — Show the reason for requesting the information and avoid dictatorial phrases.

33. *Due Date.* — It is advisable to have a set time by which the return must be in the hands of the inquirer.

34. *Duplicate Blanks.* — Send all blanks in duplicate, at least, so that the respondent may retain a copy in his files.

35. *Return Envelope.* — It is advisable to inclose a self-addressed envelope. (Use addressograph or similar device for this.)

REVIEW

1. Secure some sample questionnaires from state, national, or local administrative bodies, and test them according to the standards suggested.

2. Which of the standards enumerated seem to you to have universal application; which might be deviated from without serious results?

3. Explain and illustrate what is meant by point 17.

4. Work out alternative methods, as suggested in point 18, of arranging several questions.

EDITING OF SCHEDULES [1]

Editing is a process preliminary to tabulation. It does not necessarily imply inaccuracies in the schedule returns, although inaccuracies, some of which can be corrected by the editor, will generally be discovered in the process of editing, and in some classes of schedules as, for example, in those making returns of financial statistics of corporations or municipalities, the correction of errors by editing may materially affect the results of the tabulation. Schedule editing is, nevetheless, even in the exceptional cases noted, primarily formal rather than corrective, since the schedule data are original, and are not subject to material revision where the several replies are consistent with one another, except by referring the schedule back to the enumerating agency, or by initiating a new enumeration.

The general purposes of schedule editing are to insure, in as high a degree as possible, (1) accuracy, (2) consistency, (3) uniformity, and (4) completeness in the schedule returns.

1. *Accuracy*

Certain replies may raise a presumption of error, and in some cases this presumption may be sufficient to warrant

[1] Adapted with permission from Bailey, W. B., and Cummings, John, *Statistics*, A. C. McClurg and Co., Chicago, 1917, pp. 17–25.

investigation and verification. . . . Schedules, or copies of schedules, collected by mail from manufacturing establishments or public service corporations or steam railways, after examination in the central office, are frequently returned to the reporting agencies for correction, or letters of inquiry covering certain points in the schedule are sent out calling for correct data.

Generally, however, the editor must accept the schedule as it is presented to him without further reference to the enumerating or reporting agencies.

When inconsistent or impossible replies have been entered upon the schedule as finally accepted by the central office, it must be edited into consistency; since the process of tabulation, which follows editing, exacts absolute consistency from each schedule. This editing for consistency may be regarded as being in a sense corrective, but it is so only in a very limited and special sense, since the scope of the editor's authority to revise replies is defined in the schedule itself. All schedule replies are equally original, and the only evidence competent to justify the revision of one reply is the evidence presented in other replies. In editing for consistency the editor makes such changes only as the schedule itself demands, and he exercises judgment only in determining which of two or more inconsistent replies shall be accepted as correct. Although in some cases it may be impossible to determine with absolute certainty which reply is correct, generally it is true that a strong probability of correctness attaches to one reply, and there is the further possibility, in cases where no probability of correctness attaches to one reply rather than the other, of editing the inconsistent replies into the "no report" class.

It is extremely important that the editor should understand and observe strictly the limits upon his authority to make changes in the schedule, and it should perhaps be noted as a minor detail, first, that the editor should never make any erasures on the schedule which will obliterate the original return, and, secondly, that all revisions should be made in a distinctive ink, so that the work of the editor will always be perfectly apparent, since the work of the editor itself may be subject to revision and should in any case be perfectly distinguishable upon the schedule.

Errors subject to editorial correction in returns of financial or accounting statistics arise chiefly from misunderstandings on the part of those filling out the schedule, or from failure to make correct classifications of returns of income and expenditure in constructing balance sheets and in making up financial statements. Different practices of accounting in different concerns and in different municipalities must be reconciled so far as possible by editing. In order to avoid this difficulty the Interstate Commerce Commission has found it necessary to impose upon railroad and other corporations subject to its jurisdiction, uniform systems of accounting, prescribing in detail the accounts that shall be kept and defining precisely all items that shall enter into the capital accounts and into the income accounts. These orders of the Commission, which have been elaborated and promulgated from time to time during the past two decades, have been absolutely essential as a means of bringing in to the Commission in the annual reports from the railroad offices, data which were susceptible of tabulation. Prior to this action on the part of the Federal Commission, the various state railroad commissions had published the reports of the railroads, practically in the form in which they were made up in the several railroad offices, and these reports were so various in character that compilations of value could not be made from them. Where uniform systems of accounting have not been imposed upon corporations, schedule returns of financial data may require considerable editing.

2. *Consistency*

In editing for consistency, the first step is to determine upon a method of procedure to be followed in examining each schedule. Efficient and complete editing involves the systematic examination of all related replies in a predetermined order of examination. This sort of editing is, of course, impossible where the replies are absolutely unrelated to one another, and it is impossible as between unrelated inquiries on any schedule. It is, for example, impossible on a population schedule to check the age return against the sex return, or to check the return of nativity or of country of birth against the return of marital condition. But many inquiries are more or less interrelated, and in such cases the reply to one inquiry determines within certain limits the replies to other inquiries. Marital condition, for example, may carry certain implication as to age, since practically all married, widowed, or divorced persons are fifteen years of age or older. A native obviously cannot have been born in a foreign country — although children born of American citizens living abroad have been classified as natives of the United States in order to avoid too great detail of tabulation.

Totals which are inconsistent with constituent items shown may be entered upon a schedule, as in the case of detail of income and expenditure which does not check up with the statement of total income and expenditure; or of detail regarding individuals in a family where the total number in the family, as stated, does not correspond with the number of individuals for which returns are made; or where a family budget is incorrectly totaled and balanced.

Generally inconsistencies are evidence of carelessness on the part of the enumerator, or of misunderstanding or ignorance on the part of the person filling out the schedule.

In some cases the inconsistency is not absolute, but is of such a nature as to make the return highly improbable. The return of certain gainful occupations in the case of women and young children, for example, while it may be highly improbable, may be nevertheless within the range of possibility. It is highly improbable, but not impossible, that a child under fourteen years of age is or has been married. Generally, if the return is within the range of reasonable possibility, it must be accepted as correct unless it can be corrected by some other related reply. The return that a person was the head of a family, and was employed in some gainful occupation, together with other detail on the schedule might in some cases justify editing an inconsistent age return as "age unknown" on the strong probability that an error had been made in recording the age, possibly by omitting one figure in writing the age, as in recording a person of the age twenty years, as of the age two years.

Inconsistencies are not always apparent upon examination of individual schedules. Replies, which upon examination of individual schedules appear merely in some degree exceptional or somewhat improbable, may develop a high degree of improbability in the process of tabulation. One instance of this sort may be cited. At the census of 1900, it was found upon tabulating the returns that the number of Negroes returned as "unable to speak English" was so large as to be highly improbable. This return could not be edited out of the schedules, because it was entirely possible that any given Negro might be unable to speak English, but it was exceedingly improbable that the number unable to speak English should be so great as developed upon tabulation of the returns. Upon examinations of the schedule used at this census, the probable explanation of the erroneous returns became apparent. In contiguous columns the schedule called for answers to the inquiries as to the person's ability to read and to write and to speak English. In the case of whites, the usual and correct return to these inquiries necessitated writing "Yes, Yes, Yes," and in some cases it was "No, No, No." In the case of many illiterate Negroes, the enumerators made the partially incorrect return "No, No, No," instead of the correct return "No, No, Yes." In consequence of this accidental arrangement of columns on the schedule, the tabulation relating to ability to speak English for the Negro element had to be abandoned. At the Thirteenth Census the columns of the population schedule were rearranged, and much more accurate returns were secured to this inquiry.

In the construction of schedules it is sometimes advisable to introduce overlapping, or even duplicating inquiries, in order to provide checks for important inquiries, where the chance of error is considerable, as in the case where the inquiry calling for age is duplicated by an inquiry calling for date of birth. Inconsistent replies to such inquiries must be edited out by examination of other replies, or by an arbitrary selection of one reply as being correct. This procedure is, however, seldom justifiable, since the disadvantages of complicating the schedule more than offset any gain in accuracy in the case of individual schedules.

3. *Uniformity*

Editing for uniformity is required where replies, in themselves correct, are variously stated. Editing of occupational returns is largely of this character. A given occupation may be designated variously in different sections of the country, or it may be variously returned from each section of the country. The return may, of course, be vague and indeterminate, as where a person is returned as a "clerk" or a "mechanic" or an "engineer" or an "artist" or an "operative."

In every case it is necessary to determine upon occupational designations which will consistently group the returns for tabulation. Moreover, since the number of occupational employments returned in any extensive inquiry may amount to several thousand — at the Thirteenth Census some 9000 different employments were distinguished — and since many of these employments are each of them common to many different industries, and since occupational returns are frequently tabulated by industry as well as by occupation, some scheme of arbitrary symbols must generally be devised for editing the occupational returns into uniformity for tabulation. Commonly, the industry and the employment returned are designated by a simple combination of letters and figures, new symbols being assigned to each new employment discovered in the process of editing. The tabulation is then made mechanically from the symbols which have been edited on the schedules, in any combination that seems advisable when the editing has been completed. After tabulation the occupational designation is substituted for the symbol.

A minor instance of editing for uniformity is found in the rounding out of numbers to be stated in hundreds or thousands, instead of units, or in full units instead of in fractions of a unit. This is done where the character of the data does not warrant a statement varying by small units, or fractions.

4. *Completeness*

Editing for completeness also is formal rather than corrective. This sort of editing may consist either in entering upon the schedule derivative data, or in entering replies to inquiries which have not been answered. Not infrequently, especially in schedules calling for financial data, percentages or other derived figures are required for tabulation which are not specifically called for in the schedule. These must be computed in the statistical office and edited on the schedule. On the other hand, replies called for by the schedule may be omitted, and these must be supplied, since for purposes of tabulation a definite reply must be entered on the schedule for every inquiry calling for a reply. Where no specific reply is indicated by other data on the schedule, the reply edited in must be "no report," "unknown," or some similar entry.

REVIEW

1. Do you agree that editing is a process always preliminary to tabulation? Is not tabulation often involved in schedule making or in securing answers to schedules? How do you then support the contention of the writer?

2. What are the steps involved in editing? Do they necessarily follow the order given by the writer? Why?

3. Contrast accuracy and consistency, as developed by the writer. Are the terms used interchangeably? Do they involve the same idea? Might the data be consistent but the editor of the data be inconsistent in editing them? How is the latter condition to be guarded against?

4. Contrast accuracy, consistency, and uniformity.

5. What does the writer mean by saying that "editing for completeness also is formal rather than corrective"?

Questionnaire Relating to the Distribution, Ownership, Operation, and Physical Characteristics of Saloons, Prepared by the Chicago Commission on the Liquor Problem.

1. Give the name of owner of each saloon doing business at present, with address and police precinct.

2. State whether the saloon is controlled by a brewery, by reason of the brewery owning license to such saloon.

3. State the license record or history of each saloonkeeper, that is if such saloonkeeper has ever been in trouble or reported for violating the law; whether warnings have been given to such saloonkeeper with respect to violations or misconduct; if the license of the saloonkeeper has ever been revoked for cause; if ever convicted and fined for breaking the law; and other information of this nature.

4. State who actually operates and conducts such saloon, that is, is the man who actually operates and conducts the saloon the real owner or merely the agent or employee of some other person or party who holds or owns the license?

5. Give the name of person appearing on the city license for each saloon.

6. State number of employees of each saloon, the nature of the occupation of such employees, that is, whether employed as bartender, porter, and the like, and give name and address of each employee.

7. State whether the government liquor license is in the name of one person or corporation, and whether the city liquor license is in the name of another person or corporation.

8. State whether fixtures in saloon, as well as lease to the premises, are owned by the holder of the license, or by the person actually operating the saloon, or by the brewery.

9. State whether partitions, stalls, private winerooms, or palm and picnic gardens are permitted in and about the premises of the saloon.

10. State whether dances are permitted to be held in the rear rooms of each saloon or in any other portion of the building in which such saloon is located.

11. State whether the saloon is within 250 feet of a public or private school, church, or any public institution.

12. State whether the saloon has direct connection with hotels, bedrooms, or other private rooms, whether in the rear, side of the saloon, or overhead.

13. State whether the front, side, and rear entrances and exits to the saloon open into a street, alley, yard, or other open grounds, or otherwise.

14. State whether the saloon has a cabaret, music, or other form of amusement in or about the premises.

15. Give other facts regarding conditions in saloons not noted above.

REVIEW PROBLEMS

1. Criticize the general form of this questionnaire.
2. Using sections 9, 10, 12, 13, and 14, and following the instructions in the *Text* and in *Points to be Considered in the Use and Form of a Questionnaire*, arrange them in the form of a questionnaire, which can be statistically handled.

REVIEW PROBLEMS

1. Using the form of the questionnaire on page 239, tabulate the descriptive detail of the house in which you are living. Work out, with the other members of the class, a uniform code system to designate the presence, absence, or number of each descriptive detail.
2. Preserve your descriptions for later use.

REVIEW PROBLEMS

1. Answer question 3, Section D of the schedule on page 240 in such a form that your answer would be statistically usable for

(1) medical purposes:

(2) assignment of responsibility as between the person injured, the nature of the work done, and the condition of the machine operated.

2. Which, if any, of the questions seem to you to be poorly worded? Why?

SCHEDULE FOR DESCRIPTION OF BUILDINGS AND THEIR LOCATION.[1]

....

Dist..........Map..........Blk..........Lot..........Pg.........Line......

EXAMINED..1910

By...No...........
Assistant Assessor

Single house *one side of double house* *one of row* *Duplex house*

No..St.
Ave.

Material—*Siding drop, lap, shingles, brick, common-press, plaster, veneer, stone, cut, rough, concrete tile T. C.* Trimmings — *plain, ornamental, stone, cut, rough T. C., brick, wood.* Upon a foundation of *stone, brick, tile, concrete, posts.*

Main floor..........feet above ground.

Dimensions — *Wide, deep, wide, deep, wide, deep, wide, deep;* story high *wide, deep, wide, deep, wide, deep* story high

Projections — *One story bay window front porch* *two story bay window side porch* *three story bay window rear porch* *tower*

Roof — *Shingles, slate, tile, gravel, composition, tin, copper. Hip, gable, flat, mansard. dormers or gables.* Cornice — *plain, ornamental, wood, metal, stone, T.C.*

Divisions — *Basement, cellar, under whole, front, middle, rear* containing *storage room* *water closet* *heating plant* *laundry tubs* *bath*

1st story — *hall, parlor, sitting room, library, diningroom, kitchen, bathroom, bedroom.*

2d story — *bedroom, bathroom, other rooms.*

3d story — *bedrooms, bathroom, other rooms.*

4th story — *bedroom, bathroom, other rooms.* Attic — *rooms finished, unfinished.*

Inside Finish. Main part, lower story — *ornamental, plain, hardwood, pine, oil, paint.* Upper story — *hardwood, pine, oil, paint.*

Heating — *Stoves, furnace, hot water, steam, combination.*

Water — *Open well, city, in yard, basement, first story, second story, third story.* plumbing — *bathrooms, water closet, wash basin, laundry tray, sink, barn, — open, closed.*

Lighting — *Gas, Electric, Oil.* Fixtures — *Plain, Ornamental.*

Drainage — *Cesspool, sewer.* Building in *good, fair, bad,* repair. *Vacant, occupied, owner, tenant.* Rents at $ per month.

....
Name of *Owner, Agent, Tenant.*

$........*Rate*......$.........*per*.......$........*square*...$........*foot*........

Barn — *Wood, brick, stone, wide, deep,* stories high contains *stalls,* *living rooms*

Sidewalk — *Wood, stone, cement, brick,* Curb, *wood, stone, granite* Condition — *good, fair, bad.*

Lot Surface — *Level, uneven;* about feet *above, below* grade

Barn $............................ Bill Board..........................

[1] Taken from *First Quadrennial Assessment of Real Property of the City of Cleveland,* 1910, p. 20.

REPORT OF A PERSONAL INJURY TO AN EMPLOYEE
REPORT NO. 1

AN ANSWER SHOULD BE MADE TO EVERY QUESTION

SEC. A. EMPLOYER, PLACE AND TIME.	1. Employer's name .... 2. Office address: Street and No. .... 3. City or town .... 4. Business (state exact nature) .... 5. Location of plant where injury occurred .... Street and No. .... City or town .... 6. Date of injury .... 7. Day of week .... 8. Hour of day ....
SEC. B. INSURANCE.	1. Are you insured to provide payment to injured employees under the Workmen's Compensation Act? .... 2. If so insured, give name and business address of the insurance association or company .... 3. Has injured employee given notice in writing reserving common law rights? .... 4. If so, when? ....
SEC. C. INJURED PERSON.	1. Name of injured employee .... 2. Address .... 3. Sex .... 4. Age .... 5. Occupation when injured .... 6. In what department or branch of work? .... 7. Was this the regular occupation of employee? .... 8. If not, state regular occupation .... 9. Was injured employee piece or time worker? .... 10. Wages, or average earnings weekly ....
SEC. D. CAUSE.	1. Name of machine, tool, appliance, etc., in connection with which injury occurred .... 2. Hand feed or mechanical .... 3. Describe fully how injury occurred .... 4. Part on which injury occurred .... 5. Is it possible to provide a guard, safety appliance, or regulation in connection with this machine that might have prevented this injury? .... 6. What guard, safety appliance, or regulation to guard against the injury was in use when it occurred? ....
SEC. E. NATURE OF INJURY.	1. Part of person injured (state whether right or left in case of arms or hands) .... 2. Nature of injury, as near as possible .... 3. Attending physician or hospital where sent, name and address .... 4. State probable period of disability (number of days employee is expected to be absent from employment, dating from day of injury ....

Date of Report .... Made out by ....

3. On the supposition that you were in receipt of one hundred schedules of this type, write out a full set of instructions to a group of clerks for editing the same.

4. Respecting statistical analysis:

(1) Name a business or other problem, preferably out of your own experience, which can be studied statistically.

(2) State clearly and definitely a purpose to be accomplished in such a study of this problem.

(3) Indicate the sources of information to which you would go for data, indicating the *statistical* peculiarities, limitations, and virtues of the data.

(4) Indicate how the data would be selected, collected, or summated and what cautions would have to be observed in securing them.

(5) Define sufficiently for statistical use the units of measurements which you would employ.

(6) Formulate a *questionnaire* containing six questions bearing unmistakably on your purpose.

CHAPTER V

CLASSIFICATION — TABULAR PRESENTATION

The Purpose and Method of Tabulation[1]

Nature of Tabulation. — The general meaning of the word "table" appears to be an even flat surface with breadth not disproportionately small in comparison with length or, concretely, an object characterized by the possession of such a surface. The arrangement of ordinary reading matter is in a line or lines, while a statistical table presents itself as a surface.

The table thus differs from the ordinary page of letter type not merely in being composed mainly of figures, but also in being readable in two dimensions, that is, at least vertically as well as horizontally. "Reading matter" may also be a list of numbers. But the arrangement of the line (or "lines") of ordinary reading matter running back and forth on the page is not on a surface plan. A line of running print can be followed but one way. Such a line is like a string of beads, but with the type (as the beads) interrupted on the parts of the string extending from right to left and in position on the string as the line passes from left to right. The reader's eye must follow the string. A statistical table, on the other hand, can be read either down or across. It utilizes the dimensions of a surface. According to this conception, a list is not a table and a single column does not constitute a table.

[1] Adapted with permission from Watkins, G. P., "Theory of Statistical Tabulation," *Quarterly Publications of the American Statistical Association*, December, 1915, pp. 742–757.

A table may also sometimes be read diagonally, especially one of content and form such as to show correlation. The ages of men and of their wives, the age and the grade of school children, etc., may conveniently be compared with reference to the most frequent combinations in this way.

Matter not of a statistical character may also be put into a table when there is some advantage in reading it more than one way. Numerical data, whether statistical in character or not, are frequently best so arranged. The tabular form is used to furnish data for, and facilitate the processes of, computation, as in the familiar tables of logarithms, trigonometric functions, roots, and powers, etc., and in interest tables. Here compactness of form and ease of reference are the important considerations, but these are also the reasons for being of the statistical table. . . .

Statistical tables consist of numbers representing quantities or degrees of *concrete* things, qualities, or events. Hence the importance of statistical units and of their definite and constant significance. Indeed, the writer would describe statistics in general as concerned with concrete numbers and quantities and their relations. It constitutes a characteristic method or methods of dealing with such numbers, and also consists of the material appropriately so dealt with . . . .

Tabular presentation has conspicuous advantages as regards economy of space and of time: of space, wherever the same class designation or name is to be applied to a large number of items brought together in the table in a single line or a single column; of time, on the part of those seeking information on a specific point, in that, by using line and column as guides, the specific fact sought can be found directly. These uses of the tabular form are not peculiar to numerical tables.

Tabulation, like speech, is a device for expressing ideas,

and in particular for expressing them compactly and in a way to facilitate comparison and show relations. Ordinary linguistic symbols, arabic and other numerical notation (including the symbolic use of position), rulings and spatial relations, and sometimes forms special to tabular notation, are all employed for this purpose. As with language generally, the tabular presentation of facts should say as much as possible with a meaning as unmistakable as possible in as small a compass as possible. There should be no ambiguity; hence, for example, blanks should mean but one thing. Expression should be as direct as possible; hence, for example, information essential to a prompt grasping of the meaning of the table should not be put in footnotes if avoidable. Reasonable conventions regarding the use of symbols should be observed. . . .

Uses of a Statistical Table. — The stub of a statistical table is most conmonly a geographical classification. For groups of such classes there will usually be sub-totals which condense the more detailed classification. But the stub may consist of the names of reporting entities, as in the case of many primary tables of corporation and financial statistics. The most important statistical data for public-service corporations are usually printed in such form by the various supervising commissions, including the Interstate Commerce Commission. But for much such data, especially for the distinctively statistical as opposed to the financial part, the company unit has little significance and compilations are made by geographical or other groups of companies. Where the facts are presented by reporting entities, the tabular form may serve the purpose merely of saving space, but the totals, which are of more statistical interest, are best obtained, and their composition best shown, by way of a table. If it were possible to provide the necessary space, it would of course be best always to tabulate by such return or report units, so that the person who used the primary data could make his own groupings and combinations. However, especially where the enumeration or report unit is the individual or the private family, aggregate presentation is unavoidable. Hence the stub-items of a table represent classes, rarely also composite individuals. In publishing statistics of manufacturers and other private business enterprises, the presentation of the facts for one or few companies by themselves is expressly avoided as tending to reveal the operations of individual establishments to competitors. Such procedure on the part of the U. S. Census Bureau and the various bureaus of labor statistics is undoubtedly wise administratively, though the fact that a large business corporation with stock broadly owned cannot properly withhold from the public any sort of statistical or financial data that is of general interest should be recognized and doubtless will in time be accepted in practice. But at present only quasi-public corporations appear to be dealt with statistically according to this principle.

The statistical interest of a geographical stub is, of course, not of the highest rank. The consideration determining its use is the fact that a general or primary table is in the first instance a record and repository of data. Only to a very subordinate extent is it wise to attempt to exhibit relations and significance in such a table. In a derivative (analytical or text) table the interest is of course different. But the arrangement of the items even of a geographical stub may be made to serve the purpose of explanation where, for example, the order of magnitude or of density is followed. In the New York First District Public Service Commission reports, the arrangement of lighting companies within groups determined by intercorporate relations in the order of size (amount of revenues) somewhat increases the statistical interest of the stub, since it is a step towards making the table show correlation. It also puts first the companies in which a reader is likely to be chiefly interested, thus facilitating reference — which fact is doubtless of more practical importance than the slight aid afforded to interpretation. The order of the street-railway groups of companies in the same series of reports is in a general way that of expensiveness of line construction. These touches of correlational arrangement are suggestive of a use of tabulation which seldom affects primary tables. The correlational use, however, supposes the captions as well as the stub-items arranged according to the degree of some quality, and thus it involves cross-classification. Primary tables ought to be planned with reference to such possible use. Perhaps the presentation of such cross-classifications might well take the place of some geographical detail.

A statistical table is often merely, and always incidentally, a presentation of items going to make up a total or series of totals. The separate columns may accordingly contain things having little or no relation to each other and they may be given together merely to save space by making unnecessary the repetition of the stub. The unity of a table, however, will usually mean more than this. But it is doubtless the first or simplest purpose of a table to show this or that aggregate and how it is made up. The stub-items constitute the individual or class names for the things of which the numbers are the entries. The entries are themselves usually aggregates. But it is possible to use the tabular form for a mere tally sheet, in which case the entries represent the individual things.

In general the stub-item of a statistical table stands for a group or class of things, and the stub contains the terms of a classification. Classifications in statistics, it should be noted, must be comprehensive, hence there is usually need of an "other" or "miscellaneous" class, and commonly also of an "unknown" or "not specified" class. For the rest, all the principles conducive to right classification apply to stub and caption classifications.

It is above implied that the captions, also, as well as the stub-items, will usually constitute a classification, or perhaps more than one classification. The fact that columns commonly add across to a total column supposes this situation. The statistical table thus becomes a mode of cross-classification.

In this more highly evolved use of the tabular form, a statistical table is essentially an arrangement of numerical data by which the data are cross-classified according to two sets of terms, those of the stub and those of the captions. The device of sub-classification is also frequently introduced in the captions and stub by way of compound captions, subdivision of stub-items, and sub-totals. The more complicated classifications usually require additional tables in series.

Instead of the terms of a classification, a time series, especially a succession of years, may be used in the stub and have much the same relation to the entries, except that column totals are then not always significant. But such a table is usually derivative.

Limitations upon Tabular Presentation. — Cross-classification corresponds to what is known in algebra as combination and is covered under the topic, "Permutations and Combinations." The mathematical principle is that the number of possible different combinations of one set of things or classes of things (enumerated in the stub-items, let us say) with another set (enumerated and described in the captions) is equal to the product of the number of items in each set. This gives the number of cross-classes or entry-places in the table. There should be occasion to use most of these, or else the form of the table needs revision, or at least condensation.

The fact that cross-classification is a process of combination serves to bring out an important limitation upon the possibilities of tabular presentation. It is often desirable to show the associations or combinations of the units under three classifications or sets of cases. If the third of these classifications is merely twofold, the space required is merely double what it was before. If there are 12 rubrics under the third classification, the normal requirement is for 12 times as much place, or probably 13 times as much, since a total of the 12 classes will be desirable. If the original stub provides for 30 items and there are 10 columns, a presentation of all the possible combinations with a further series of 12 classes will require $30 \times 10 \times 12$, or 3600 cross-classes or entry-places.

If it is desired to show completely by tabulation the relations between nativity in 12 classes, age in 10 classes, sex in 2 classes, residence in 50 classes, and occupation in 100 classes, supposing every possible combination will require an entry-place, the number of cross-classes will be $12 \times 10 \times 2 \times 50 \times 100$, or 1,200,000. If the 50 residence rubrics are made the items of the stub and 10 columns may be put on a page, that would mean 500 entry-places to a page. The presentation of the facts would, therefore, require 2400 pages. But the number of rubrics under each classification is fewer than it might be desirable to use. The above computation, moreover, does not provide for totals. Of course, much space could in practice be saved by reason of the omission of provision for impossible or infrequent combinations. Young children, for example, will not be found in occupations. However, the limitations upon what we may call *complete tabulation* are evident. The size of census volumes, even with their limitations, is thus explained.

The difficulty in question is avoided by seldom attempting complete tabulation. Some of the combinations are not important or not of special interest. The classification of those in a specific occupation by nativity, for example, is of interest for comparatively few occupations and comparatively few localities. It may often be assumed that the variation within one kind of classification in terms of another classification will be so small that a presentation of the facts for all of the first class combined will sufficiently meet ordinary statistical requirements. Detailed compilations also may often be made to serve for a number of years, provided the proportions found are representative and quite constant. The frequent necessity of resorting to such methods — the necessity in particular of using alternative classification instead of cross-classification — explains why a given statistical compilation will seldom enable one to answer all the questions for which a solution is sought. The facts are contained in the returns but they cannot all be presented.

A report schedule from which tabulations are made is commonly itself in tabular form and may contain a cross-classification. Only one who has had practical experience with the problem of devising a general table or tables to contain what is most important in such returns can appreciate the difficulty of obtaining satisfactory results in a limited space. But the reader is prepared for an application of the theory of mathematical combinations to such a case. If only 50 such report schedules are to be tabulated in a way to show the individual returns and supposing the schedule has 10 stub-items and 20 captions, then in order to present *all* the facts it would be necessary to provide at least 200 columns of 50-line tabular matter. Alternative tabulation, on the other hand, which would utilize only the cross and down totals of the schedule, would require 30 columns. It is assumed, of course, that the data of each schedule are themselves aggregates and that each such aggregation has interest of its own. If only the totals for the 50 returns taken together are wanted, only as many entry-places are required as are contained on one of the schedules, that is, $20 \times 10 + 31$ (for totals), or 231 in all — which is a table of modest dimensions. Enumeration schedules, it should be noted, are not often of a character to raise this question in just this form. . . .

With our present-day mechanical facilities for "tabulation," the process of subdivision and cross-classification of aggregates is limited rather by the degree of significance of the results, and by the cost and awkwardness of voluminous reports, than by the time required to make the necessary sortings and counts of cards already punched. While the mathematical theory of combination is a good point of departure in planning tables, most combinations of the terms of diverse classifications, even if they occur, have no concrete significance.

Comprehensiveness, Comparability, and Compactness as Essentials of Good Statistical Tables. — The significance of a statistical table, as of statistics generally, depends very largely upon its being comprehensive for the field it covers. Truth in its statistical aspect is representativeness. The only absolute guaranty of the representative quality of an aggregate is that it reflects all the units within its scope. According to the mathematical theory of probabilities, much less is necessary, but this theory does not take account of the selective tendency of events and of observation, for which the statistician must be continually on his guard. The point is illustrated by the well-known difference in quality between results obtained by complete enumeration and those obtained from a circular letter or questionnaire.

A table should not be composed of mere samples. It is better to make it of narrow scope but comprehensive as far as it goes, *i.e.* within its territorial or other limits. A table, furthermore, is likely to be one of a series, which should all be on the same basis, or, at least, conform sufficiently to the basis of the series so that its representative quality and the comparability of its totals are not appreciably impaired. The most surely understood uniform basis, meeting all the requirements of comparability, is the comprehensive basis. When a table falls short of the basis of its fellows, but in a way not such as to compel its omission altogether, the appropriate place to indicate what is lacking is a general note. Sometimes it may be well to have two sets of totals to a table, one on the most comprehensive basis, and one less comprehensive, but such as to supply aggregates for data that, though falling short of perfect comprehensiveness, may be of qualified value in other ways, as for example, in the computing of ratios. On the other hand, if it is desirable to present information in connection with only one of a series of tables, it is well, in order to avoid impairing the comparability of one table with the others of the series, to put the data that exceed the standard scope in brackets and not take them into the totals, thus letting them be *in* the table for purposes of reference, but not strictly *of* it. Uniform comprehensiveness upon some definable basis is the ideal standard. Even a small per cent impairment of comprehensiveness may mean a large decrease in tabular efficiency.

The same principle applies with reference to corresponding tables for a series of years. While it is desirable that new data be made use of, full notice of a change of basis should be given and it is often well to give figures and make comparisons on both the old and the new basis for the first year of the change. Especially in derivative tables attention to comparability is imperative, without regard to cost in the way of added complexity, etc. Ratios, for example, should usually be given on both bases where there is a change. This again is a question of representativeness, though here differences between aggregates, rather than the aggregates themselves, are under consideration. How important this question is in another of its phases is illustrated by the place commonly given to averages, *i.e.* representative numbers, as the gist, if not the substance, of statistics.

The complement of the requirement of comprehensiveness is that of compactness. It is of the essence of a table to convey a large amount of information in a small space. Hence sparsely tenanted columns are an eyesore, and blank columns, even where the original classification may have reasonably planned to use them, should not be tolerated. Blank lines are hardly less justifiable. Classifications should be revised when the data as spread out show such waste of space. Unrepresented classes may be disposed of in the notes. Sparsely tenanted columns should be consolidated, subdivisions of entries being indicated by footnotes if desirable. A "miscellaneous" column may often be employed with reference to such residual classes. It should never include more than a small per cent of the material of the table. But sometimes the desirability of keeping up tables on a uniform plan, *e.g.* through a series of years, may justify continuing sparse columns till a comprehensive overhauling of the form of tables is undertaken.

The table must ordinarily be planned with reference to fitting the printed page, as single-page lengthwise, single-page upright, twin upright, or as a series of such. Hence dimensions in terms of columns and lines must often be carefully studied before being finally fixed. The large page and the resulting unwieldy size of most statistical volumes are due to the need of space for manœuvering the tabular matter. Often the presentation in sections of what is functionally one table becomes necessary.

General Tables and Derivative Tables Distinguished. — A table serving primarily the purpose of a repository of comprehensive statistical data is distinguished as a general table, also, with reference to its being closest to the original data, as a primary table.

Derivative tables are summaries and auxiliary ratio tables. They may be usually distinguished as text or analysis tables. But some ratio tables, or at least some ratios, are often included among general tables. Derivative tables are based upon general tables and contain matter suitable for incorporation in analysis. They may vary in form from year to year according to the exigencies of the situation and according to the points emphasized in the text. Unlike the general tables they will usually contain data and comparisons, including absolute and per cent increases, for several years. Just as general tables serve to show in terms of absolute numbers the composition of aggregates, a derivative table frequently serves the purposes of explanation correspondingly by means of per cent distribution. If text tables contain data taken direct from returns, these are so treated because of lack of comprehensiveness in the data, or of perennial interest in that kind of data. Explanatory and qualifying statements contained in general-table footnotes should, unless unimportant, be either repeated or referred to in footnotes, or in text immediately adjacent to the text tables.

It is the common practice of statistical bureaus to number tables serially for each report. If Roman numerals are used for the general tables, arabic numerals are used for derivative tables, or vice versa. . . .

No strict line can be, or need be, drawn between what

should go into general and what into text tables, though the fact that ratios are logically a part of the analysis gives the analytical text, if there is any such, a strong claim upon them. Grand totals certainly go with the general tables not only as closing them up but also because of their importance as a proof check. But divisional totals serving the purpose of a summary may go in either place. Ratios, too, may come to have so thoroughly well established a place as to be in effect a part of the data that the public will expect to find in connection with the general tables. A derivative table in a report containing the corresponding primary tables is seldom to be considered a thing by itself to the extent of requiring no reference to its sources on the part of a reader who uses it carefully.

Comparisons with previous years — or with corresponding months (or other portions) of previous years — are also strictly a part of analysis, but their significance is so direct and their meaning in general so unmistakable that some of them may well be looked for in the general tables. They are made much of especially in commercial and financial statistics. The United States Census is liberal in presenting comparisons for previous decennial years in its general tables.

General or primary tables rightly occupy the largest place in most government statistical publications. Indeed, some official statisticians feel that the preparation and presentation of the primary tables is their whole duty. But some working-over of the raw material by those directly concerned with its compilation is desirable, if for no other reason than the beneficial reaction on the original data and tables consequent upon analyzing and applying them to the solution of scientific and practical problems. Proper emphasis upon the function of such statistical publications as sources does not preclude brief suggestive analysis, in addition to the necessary descriptive and cautionary remarks.

The Rounding and Abbreviation of Numbers. — The use of rounded or cut-off numbers should seldom be adopted in general or primary tables, though doubtless desirable in derivative or interpretative tables. The practice is often recommended without reference to, or due emphasis upon, this very necessary qualification.

Even in derivative tables, the giving of a large number, for example, millions of inhabitants, to the last digit would mislead by its supposed suggestion of "spurious accuracy" only in the case of a reader who would have at least equal difficulty in understanding what the rounding of the figures meant. The notion that we should print numbers showing the digits only in so far as they are known to be accurate, or on the basis of the theory of probabilities considered to be so, is impractical to the height of absurdity. The truth of the stated population of New York City — 4,766,883 in 1910 — is not of a nature to imply that the figure 3 in the units place has statistical significance. The statistician knows that the last four digits are neither more nor less accurate or truthful if made to read 7000 instead of 6883. He does not need to be reminded that the 117 has no objective or exact meaning in such an aggregate. It is seldom necessary to indicate that large numerical aggregates are approximate as to the right-hand figures.

But there is also a positive objection to the rounding of such numbers. From the point of view of statistical administration it is important that, for example, the population of a large area be the total for all its parts down to the smallest district for which separate figures are given, some of which in the instance referred to actually have less than 117 inhabitants. Rounding an absolute number is never obligatory and should never be done in a way to deprive any one of the possibility of completely checking the number and of using for this purpose, if for no other, the unmodified original aggregate. Primary numerical data should not be rounded.

As regards ratios, too, their mechanical computation with equal ease to a larger as to a smaller number of places makes the decision of how far they should be carried a question of conventional expectations and of economy of attention rather than anything more fundamental. This statement does not refer to (and does not apply for) slide-rule computations. The carrying out of ratios to two decimal places (or for per cent to hundredths of one per cent) seems to be the most satisfactory practice for most cases, so far as fractions are desirable, though only the first place will usually be itself significant, the second serving rather to qualify the first. Where three decimal places are used, the printer, and sometimes the reader, will easily mistake the point for a comma.

But much depends on how far it is the statistician's aim to make his material popular — an end that is, of course, entirely worthy in itself. The desirability of rounded and abbreviated numbers, also of the use of few numbers, in statistical exposition is chiefly of the same nature as are the claims of stylistic elegance or of force (as a writer may prefer or the conditions require) in the use of the English language. The first duty of one presenting statistical results is to be adequate and accurate; if possible it is well for him to be also elegant, or forcible, or whatever else may be desirable, in his choice of words and of numerical expressions.

The process of rounding or cutting off numbers is by no means simple or a matter of course. On the contrary, it requires considerable statistical technique — else totals will be found not to check with items and ratios not with the data from which they are derived. It may be noted incidentally that where it may seem desirable, as frequently in the case of estimates, to round or abbreviate both a relative number and the corresponding absolute number, one cannot do both and at the same time preserve the requisite verifiable relation between the two. This fact counts against the rounding even of estimates, though some sign of approximation is in such cases especially desirable.

Tabular Notation. — The rounding and abbreviation of numbers is strictly a part of the subject of tabular notation, but so fundamental as to affect the character of the statistical table as such. The word "notation" properly refers to the relation between the signs and symbols used to convey the meaning of any part of the table and the significance arbitrarily or conventionally attaching to them. To illustrate, it would seem that the last two digits, 83, of the figure for the population of New York City in 1910, preceded as they are by five other digits having the significance of position proper to them according to the arabic numerical notation, ought, without difficulty, to be interpreted as having a different statistical significance from the figure 83 as arrived at, for example, by a careful housewife on inventorying her pieces of silverware preparatory to putting them into safe deposit, or by a dairyman counting his stock.

The signs used in tabulation are chiefly arabic numerals and the letters of the alphabet in their various appropriate combinations. The position of such a sign may be a part of the notation. The notation of a table is the language in which its import is expressed; and that language should be as direct, concise, and unambiguous as it is possible to make it.

The technique of statistical notation has not reached a

high stage of development. The writer, at any rate, feels that the tendency among statisticians to treat a table as a mere repository of numbers and to indicate in footnotes any state of facts not so represented is objectionable. The absence of a report, the failure to segregate returns, the character of an entry as estimated or as incomplete — all these are matters that can be shown by appropriate signs on the face of the table. The best policy would seem to be to make the tabular entries self-explanatory to as high a degree as possible, for the purposes of the particular tabulation, by the use of word or other non-numerical sign entries where feasible. Footnotes are thus reserved to supplement or qualify both numerical and sign entries and especially are not intended to take the place of lacking numbers. But the technique of tabular notation lies outside the scope of a discussion of the general aspects of statistical tabulation.

REVIEW

1. Why may a statistical table be spoken of as a "surface"? From what angles may such a surface be viewed?

2. Contrast caption- and stub-headings. May they always be interchanged? Why? Work out a "treble" table, and interchange the headings. What is the result? What conditions control the order of items in both?

3. Formulate a general statement showing the "Limitations upon Tabular Presentation." How are these overcome?

4. Why may "comprehensiveness, comparability, and compactness" be held to be essentials of statistical tables?

5. Contrast general and derivative tables.

6. How is the practice of rounding and abbreviating numbers in tabulation related to accuracy, to "spurious accuracy," to compensation of errors, to the serviceability of tables?

Standardization of the Construction of Statistical Tables[1]

The progress of every art should be marked by the accumulation of an increasing stock of generally accepted practices. As these practices obtain common approval, they should be recognized as standard and regularly followed until more satisfactory methods are discovered. A measure of standardization is thus a normal feature of development.

Standardization of statistical practices should not be invited, however, without recognition of its dangers. Like "law and order" in civil life, standardization may easily be overdone. There is always the risk of formalism. But kept within proper limits, standardization has a steadying influence which tends to accelerate, not retard, the improvement of statistical exposition. It effects good order, and is an unmistakable mark of real progress.

It is consequently profitable to consider from time to time the extent to which standardization can advantageously be accepted. In statistical exposition, the standardization of graphic methods has been one of the gratifying advances of recent years. To what extent has there been and to what extent are there further opportunities for a similar standardization of practice in the methods of tabular presentation?

In considering this question, it should not be thought that standardization is accomplished only through the conscious adoption of rules and regulations set up by

[1] Taken with permission from Day, Edmund E., "Standardization of the Construction of Statistical Tables," a paper read at the Eighty-first Annual Meeting of the American Statistical Association, Chicago, December, 1919, and later published in revised form in the *Quarterly Publications of the American Statistical Association*, March, 1920, pp. 59–66.

recognized organs of authority. Standardized statistical practices may evolve by imperceptible degrees through the influences of imitation and prestige. This is particularly the case if some one statistical bureau is the fountain-head of governmental practice. The working rules of such an office tend to become the rules of a following of less influential practitioners. Standardization of this kind is going on at all times. Such standardization of practice as we have to-day in statistical work in this country is almost altogether the result of the influences of imitation and prestige.

Unconscious standardization of this sort has already made substantial progress with regard to the structure of statistical tables. Without attempting a complete enumeration of the rules observed by competent authorities, a few of the standard practices may be noted in passing. Thus it is generally recognized: (1) that every table should be self-sufficing, containing within itself a clear explanation of the meaning of the items displayed; (2) that every table should be logically a unit, containing only data which are intimately related with one another; (3) that column- and row-headings should be brief, unambiguous, and self-explanatory, table footnotes being used when necessary to make the headings perfectly clear; (4) that coördinate and subordinate relationships among the column- and row-headings should be shown by variations of boxing in the captions and of indentation in the stub; (5) that varieties of letters, figures, lines, column-widths, and interlinear spacings should be employed to facilitate easy and intelligent use of the table; (6) that columns and rows should be lettered or numbered if cross reference is desirable; and (7) that sources and units should invariably be indicated. The common acceptance of these principles represents no mean advance in the standardization of statistical table structure.

It is to be observed, however, that the standardization thus far effected concerns primarily the constituent parts of the table, not the table's general form. The choice of position between columns and rows, the arrangement of the several columns or the several rows, and the location of particular columns toward the left of the table or of particular rows toward the top, seem still to be matters of individual preference, if not of chance. It is important to consider how far standardization of the general form of statistical tables is feasible and desirable.

Standardization of the general form of statistical tables must begin with a distinction between general-purpose and special-purpose tables. The general-purpose table is designed to bring together in most convenient and accessible form all the data bearing upon a given topic. The special-purpose table is intended to throw into relief relationships of special significance in a given study. The general-purpose table is an orderly presentation of statistical material; the special-purpose table, a record of the results of statistical analysis. Of course, a measure of analysis is a prerequisite even of the general-purpose table, but the analysis is of a different order. It is the analysis essential to effective enumeration and tabulation, not the analysis accompanying specific interpretation. The analysis required for the special-purpose table is directed toward a particular issue. The problems of good table structure are essentially different for the two types of tables.

Since the construction of the general-purpose table is the simpler case, it first will be examined. In considerable measure, the general-purpose, or primary, table is a creature of the physical form of the medium in which it appears.

Upon the one hand, the table tends to expand to accommodate the large body of data pressing for inclusion. Upon the other hand, the capacity of the printed page — even if it be folio — stands as a limit on the indefinite enlargement of the table. Tables which are allowed to exceed the dimensions of the page and have to be folded in are everywhere recognized as objectionable. Loose tables, separately printed in large irregular sizes, are as bad, if not worse. Tables running across two pages facing one another are reasonably satisfactory but are to be avoided where possible. Tables which are presented at right angles to the text fall into the same class. In general, the single page, held as when reading the text, is the maximum size to which the statistical table should be permitted to run. Primary tables usually press upon this physical limit; their outside dimensions are thus independently determined.

Within the table, similar influences are at work. Whether given arrays of data shall be exhibited in columns or in rows is commonly a question of the difference in the vertical and horizontal capacity of the page. The maximum number of lines in a table is several times greater than the maximum number of columns. Consequently the arrays having the greatest number of items are naturally assigned to the columns, the other arrays to the rows. Once a given set of headings has appeared in caption- or stub-position, there is a strong presumption in favor of its occupying the same position in other related tables, for the transcription of data from general tables is thereby facilitated. Upon the whole, however, the assignment of columns and rows rests fundamentally upon the greater capacity of the column: a factor not subject to modification by the statistician.

A much larger measure of option may be exercised in fixing, in a general-purpose table, the order of columns and of rows. Almost any systematic plan may be adopted; but the most satisfactory arrangements are the alphabetical, chronological, geographical, or according to the magnitude of the items. There are no grounds for urging the adoption of any one or two of these arrangements to the exclusion of the others. Now one best serves; now another. One rule, however, should govern the final selection in all cases: that order should be employed which keeps the details of the table most generally accessible. Readers will come to the table with a variety of interests. They should be given that table from which in general they can most easily draw the information they seek. Arrangement according to magnitude or importance of items is less satisfactory in general-purpose, than in special-purpose, tables, because it depends upon analysis from a single point of view and it is frequently unwise to commit the table to this particular viewpoint. The other arrangements better meet the variety of needs which a primary table is designed to serve. The important end is to secure some logically and commonly understood arrangement which opens the table to easy transcription.

When geographical or chronological orders are adopted, a decision has to be reached as to what items to place at the top and left and what items at the bottom and right. In the tabular arrangement of the states of this country the grouping and order followed by the Bureau of the Census may be recognized as standard; the northern New England states stand at the head of the list, the southern Pacific states at the foot. In general, the best statistical practice for this country would seem to run geographical series from north to south and from east to west. With chronological series the case is not so clear. Upon the whole, however, for general-purpose tables, the Census Bureau practice of placing most recent dates at the top and left seems commendable if there is a fair presumption that the figures of most recent date will be most frequently transcribed. When, however, the data will probably be transcribed in entirety as time series it would seem preferable to place the figures for earlier dates toward the top and left. The rule to apply in all these cases is simple: the most generally useful data should be located toward the top and left where accurate transcription is rendered easier by close proximity to the column- and row-headings.

The general or primary table exhibits no specific analysis. Its form is in considerable measure the resultant of the physical limitations of the page and the necessity of presenting a maximum body of data in a way to make the most generally useful parts most readily accessible. The derived or analytical table is a different statistical device. A derived table is essentially deficient if it fails to exhibit a carefully formulated analysis. It should be constructed to assist a specific interpretation; every effort should be made to make the table simple; it should contain only those items valuable to the analysis, arranged so as to encourage the deductions the reader is expected to draw. If any line is to be drawn between statistical tabulation and statistical analysis, the primary table displays the results of tabulation, the derived table the results of analysis.

Despite this fundamental distinction between primary and derived tables, it is to be admitted in the first place that the derived table is not altogether free from the influences of format which plays so important a part in shaping the primary table. For example, if the number of subdivisions in one classification of an analysis is much greater than in the other, it may be necessary to put the more extended classification in the stub simply because stub-capacity is normally so much greater than caption-capacity. Similarly, if the designations in one classification are much longer than in the other, it may be necessary to place the classification with longer headings in the stub, since neither of the alternatives — printing the longer headings vertically at the top of the columns, or widening the columns to accommodate the longer headings horizontally — is at all satisfactory. Such crass considerations as these are at times decisive in determining the structure even of the derived table. But they play a much less important part with the derived table than with the primary table. As a rule the statistician is able to make the general form of the derived table serve the exposition in hand.

One of the most fundamental questions of structure is the assignment of data to columns in some instances, to rows in others. This matter should be settled in the derived table with reference to what comparisons it is most important to present. Comparison of like items in a column is much easier than of like items in a row. It is believed that recognition of this fact will commonly throw chronological, geographical, and quantitative classifications into the stub, qualitative classifications into the caption; but this is not a necessary outcome. The important principle is to use the column position to promote the more significant comparison.

Arrangement of the several columns and of the several rows in the derived table will be determined by the particular character of the analysis in connection with which the table is employed. If the analysis is of a temporary distribution, a chronological order will be adopted; if of a spatial distribution, a geographical order. If the items are component parts of an aggregate, arrangement will be either according to the relative magnitude or importance of the item, or according to some other order generally recognized in the analysis of the data in question. Presumably the alphabetical arrangement will seldom be followed, since it does not directly disclose significant relationships. Ordinarily the purpose of the analysis will indicate clearly enough the order in which the columns or the rows should be placed.

Naturally the arrangement of columns and rows should give proper regard to the fact that the most conspicuous position in a statistical table is at the top and left. While it is generally true that derived tables are designed to bring out relationships rather than individual items and that these relationships are properties of the table as a whole rather than of particular parts, it may be desirable in some tables to focus attention especially upon certain more important items. When other considerations will permit, these more important items should be placed in the most exposed positions of the table: namely, at the top and left next to the captions and stub. This rule is a sufficient warrant for placing totals at the top and left when they are clearly the most significant items of the tabulation, and when placing them at the top and left will not give serious offense to the users of the table. If either of these conditions is not present it would seem preferable to place totals in the positions in which most readers expect to find them, namely, at the bottom and right. There appears to be no adequate reason for departing from the established practice of reading time from top to bottom and left to right. In derived tables, figures for later dates should appear toward the bottom and right. It is the relation between items, not the individual item, which is significant in time series. For many reasons we are accustomed to thinking of the upper or left-hand of two figures as being the earlier, and we draw our conclusions accordingly. Furthermore, this rule is already thoroughly incorporated in our graphic practices. To have diametrically different rules for graphic and tabular presentation would be unfortunate. The Census Bureau practice of placing data for most recent dates at the top and left is therefore not to be approved for the derived table. Effective exposition of the statistical evidences is better served by the order which seems most natural to the great majority of readers. Arrangements of columns and rows should hold fast to the purpose of facilitating interpretation.

If the dominant purpose of the derived table be kept in mind, many problems of tabular arrangement will be readily solved. Percentage distributions will be placed next to the corresponding absolute figures or in a separate portion of the table according to the emphasis of the analysis. To facilitate comparisons of relationship, the arrangements adopted in one table of an analysis will be followed as closely in the other tables as other more important considerations will permit. Columns and rows which are to be compared with one another will be brought as closely together as possible. Unnecessary digits will be dropped and items given in round numbers to simplify the presentation. The aim throughout will be to make the derived table an effective instrument of statistical exposition.

If such are the considerations involved in the construction of statistical tables, what conclusions are to be drawn regarding the possibilities of standardization of table structure? Upon the whole, the opportunities for complete standardization seem slight except with regard to the elements from which the table is to be constructed, and certain lesser matters of general arrangement. More is to be gained at this time from a clear recognition of important guiding principles in table construction. Careful attention must be paid to the difference of purpose in primary and derived tables. The primary table must be made to offer its items for easy transcription; the derived table, for ready deduction. If statistical tables are formed with nice regard for those fundamental aims of tabular presentation, standardization may well be allowed to proceed as it has heretofore through imitation of the most satisfactory existing practices. Untiring experiment with varying forms and ready acceptance of improvements are for the present the most promising means of securing better construction of statistical tables.

REVIEW

1. In the discussion of the size of general-purpose tables, what use of the tables has the author in mind? Would you support his contention respecting such tables when they are prepared for office use only? What criteria on size would *you* set up for this use?
2. Can stub- and caption-headings be interchanged with equally good results, assuming that the page will comfortably admit of either arrangement? Make such a change, using the outlines of single, double, and triple tables. What is the effect in each case?
3. Do you agree with the author's statement that "almost any systematic plan may be adopted" . . . "in fixing in a general-purpose table, the order of columns and rows?" Compare this generalization with the contentions in the *Text*.
4. Can a line be drawn between "statistical tabulation and statistical analysis"? What answer would the *Text* give to this question?

Statistical Standards in Tabulating Facts [1]

Tabulation is a means, *first*, of recording in fixed form a classification previously developed, or *second*, of placing

[1] Adapted from Secrist, Horace, "Statistical Standards in Business Research," *Quarterly Publications, American Statistical Association*, March, 1920, pp. 53–54.

similar facts into juxtaposition or into groups as a preliminary to a final classification. It is a device for projecting on a surface, capable of being read in two dimensions, a classification which has been worked, or is being worked, out. It is a method of recording a process of thought. It is inelastic in structure; the facts which it contains are in truth "locked up." Classification precedes, tabulation follows. The sequence of thought is from purpose to method.

The statistical standards to which tabulation must conform are as follows. It is necessary to say that it is not my intention so much to formulate a set of rules governing the make-up of tabulation forms as it is to develop statistical standards in tabulation of permanent value, the realization of which may require a variable technique.

First. — Every tabulation surface should faithfully record the classification which it is intended to depict. The purpose of tabulation and the standard to which it must conform cannot be divorced.

Second. — There is always a *best* form of tabulation for a given purpose, as there is a most logical basis of classification. Indiscriminate choice of forms is as much without justification as is a meaningless or superficial classification.

Third. — Every tabulation should be adjusted in form and complexity (*a*) to the subject matter which is to be expressed, and (*b*) to the person for whom it is prepared or the end to which it is addressed.

Fourth. — The order of detail in tabulation forms should be adjusted so as to be emphatic. It should be natural, not artificial; convincing, not purposeless.

Fifth. — Statistical tables should carry only relevant data. The reciprocal relation between relevancy of fact and the purpose to be accomplished by tabulation is the thought which is stressed.

Sixth. — Statistical tables should carry on their face both their justification and their explanation.

Seventh. — The details of statistical tables should be mechanically accurate and their grouping and arrangement consistent, logical, and serviceable.

Eighth. — The natural order in classification is from detail to summary; the serviceable order in tabulation is from summary to detail.

Ninth. — Brevity is said to be "the soul of wit." It is equally true that conciseness in tabulation is the secret of its effectiveness for most practical purposes.

A CENSUS CARD

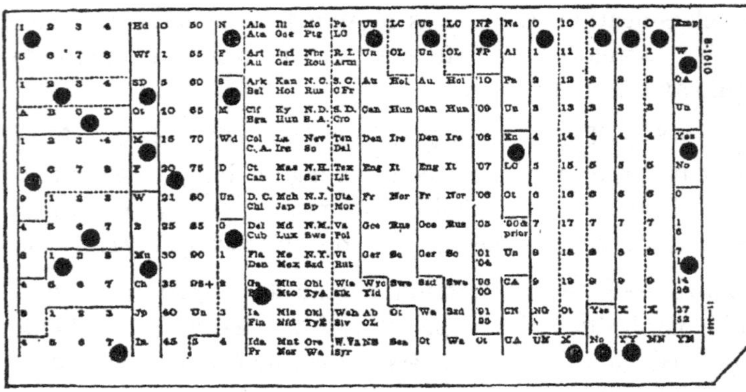

This illustration shows one of the 92,000,000 cards used in tabulating the population returns at the census of 1910. The holes in the four numbered spaces at the left are arbitrary symbols indicating the state and district in which the person to whom the card relates was enumerated; those in the other "fields" describe his characteristics. Thus, the person to whom this card refers resided in enumeration district No. 924 (Maynard, Middlesex County), state of Massachusetts; was a son of the head of the family in which he lived; mulatto; 20 years of age; native born; single; born in Georgia; father born in United States; mother born in United States; spoke English; was an agricultural laborer; was out of employment on April 15, 1910; was out of employment between 7 and 13 weeks in 1909; could read and write; did not attend school; and was not a veteran of the Civil War.

REVIEW PROBLEMS

Tabulation

1. Secure some blank Hollerith Tabulating Machine Cards. Using the detail provided on the schedule form p. 239, showing descriptive detail of your house, draft a Hollerith card form which could be used in tabulating the data.

2. Draw up three box tabulation forms for the detail of this schedule so as to show the relation of the size of the houses (1) to the number of rooms, (2) to the type of heating equipment, (3) to the lighting equipment. Give each table a suitable title, and prepare the forms in conformity with the discussion in the *Text* and *Readings*. If these conflict, choose the form which best suits your purpose and justify your method.

3. The following data in relation to registration at Northwestern during the second and third terms, 1918–1919, are to be tabulated so as to compare (1) men and women, (2) time of withdrawals, (3) source of registrants in the third term. Follow the suggestions in Chapter V of the *Text* relative to the make-up of tables. Give the table a suitable title.

Registrants 2d term 1918–1919, men, 536, women, 844. Withdrawals during the 2d term, 1918–1919, men, during the term 31, at the end of the term, 37; women during the term, 37, at end of term 68. Registrants 3d term: men, from 2d term 458, former students 51, new students 40; women, from 2d term 739, former students 19, new students 34.

4. The following types of data relative to employees at each of two establishments, "A" and "B," are available.

(1) Length of service — expressed in weeks (length of service groups).

(2) Type of occupations — laborers and operatives.

(3) Number on pay roll at end of year.

(4) Number separated during the year.

a. Draw up a table *form* so that those on the pay roll at the end of the year may be compared directly with those who separated during the year for each type of occupation for each of the establishments. (Use the length of service groups as the stub.)

b. Draw up a table *form* so that the laborers and operatives on the pay rolls at each of the establishments may be compared with those who separated during the year. (Use the length of service groups as the stub.)

c. Draw up a table *form* so that the two establishments may be directly compared for each type of occupation, for those on the pay roll at the end of the year and for those who separated during the year. (Use the length of service groups as the stub.)

5. Using the following tabulation of Failures in the United States, write a descriptive comparison, two hundred words, of the conditions in 1919 compared with 1918.

In what ways, if at all, is the discussion of the advantages of tabulation, *Text* pp. 119–125, borne out?

SUMMARY — UNITED STATES [1]

Failures due to	Number		Assets		Liabilities	
	1918	1919	1918	1919	1918	1919
Total	5515	9331	$55,361,296	$70,322,293	$115,549,659	$137,907,644
Incompetence	2109	3409	$11,730,114	$20,967,819	$26,068,530	$37,139,453
Inexperience	307	629	1,740,312	2,919,880	5,510,902	6,508,802
Lack of capital	1669	3093	15,837,726	20,516,528	29,378,542	42,543,457
Unwise credits	72	123	2,869,310	971,439	4,534,615	2,436,522
Failures of others	97	86	2,046,947	2,785,374	3,844,066	4,558,718
Extravagance	59	56	612,889	389,004	1,374,864	827,083
Neglect	93	139	340,426	456,907	934,622	1,178,563
Competition	59	116	476,852	592,176	945,009	1,045,733
Specific conditions	623	1107	12,095,267	13,779,286	23,671,566	27,312,198
Speculation	37	33	1,112,845	884,453	2,640,534	1,668,649
Fraud	390	540	6,498,608	6,059,427	16,646,409	12,688,466

[1] Bradstreet's, January 31, 1920, p. 81.

CHAPTER VI

DIAGRAMMATIC AND GRAPHIC PRESENTATION

Rules for Diagrammatic Presentation of Statistical Data [1]

A. General Make-up of Diagrams

1. *Data to accompany diagrams:*

The data shown graphically in a diagram should be given in tabular form either beside or within the diagram, or in close proximity in the text. Care should be exercised, however, to place figures so as not to disturb or distort the visual impressions conveyed by the chart.

2. *Scale units:*

In general, in the laying off of scales, the scale intervals on any single diagram should be exactly proportionate to the gradations of number, size, or time represented (the logarithmic scale constitutes an exception to this rule). . . .

3. *Scale figures:*

Figures for the scales of a diagram should be placed at the left and at the bottom or along the respective axes. . . .

[1] Taken from Day, Reed, and Secrist, "Rules for Graphic Presentation of Statistical Data," in *Weekly Statistical News*, Central Bureau of Planning and Statistics, No. 5, Oct. 10, 1918, Washington, D. C.

4. *Base lines:*

   It is well to distinguish — as by heavier inking — lines which represent standards of attainment or bases of measurement or comparison. . . .

5. *Arrangement of items:*

   Items should be grouped so as to facilitate the comparison of items most significantly related. Within groups, some systematic order should be adopted. The most serviceable arrangements are according to (*a*) the sequence of the items in time, with the earliest at the left; or, (*b*) the size of the items, with the largest at the top or at the left; or, (*c*) the favorableness of the items, with the most favorable at the top or at the left.

6. *Position of titles, etc.:*

   So far as practicable, all printing upon a diagram should be so placed as to read with ease from the bottom of the sheet.

7. *Use of colors:*

   Where a need for duplicates may arise, charts should be made entirely in black and white. The use of colors is not recommended except for large wall charts.

8. *Size of sheet:*

   Avoid irregular sizes of paper. As far as practicable, the established correspondence sizes ($8 \times 10\frac{1}{2}$ or $8\frac{1}{2} \times 11$) are to be used.

B. Choice of Graphic Forms

1. *For simple comparisons of size:*

   a — *Bars* — Bars are the most satisfactory graphic device for this purpose. In general, all the bars used in the diagrams of a single study should be of uniform width. . . .

b — *Lines* — When a large number of separate items have to be shown in a single diagram, lines may be employed in place of bars.

c — *Position* — Bars (or lines) are best placed horizontally. . . .

2. *For comparisons of component parts:*

a — *Subdivided bars* — Subdivided bars are the most satisfactory form for this case. . . .

b — *Cross-hatching* — Cross-hatching is the best way in which to distinguish the component parts. . . .

c — *Position* — Horizontal bars are to be preferred to vertical, except when the items are separated by intervals of time, in which case vertical bars should be used. . . .

3. *For displaying frequency distributions:*

a — *Vertical columns (histogram an alternative)* — In general, the vertical bar (or column) form is to be used. The straight-line histogram, however, is a satisfactory alternative.

b — *Position of scales* — the scale for the variable is to be placed along the horizontal axis; the scale for the frequencies along the vertical axis.

4. *For showing geographic variations:*

a — *Dot maps* — Where the variable takes the form of varying numbers of a given item, the situation is best represented by a dot map in which each dot represents a fixed number of cases. All the dots should be of uniform size and should be evenly spaced over the areas in which the actual items have appeared.

b — *Shaded maps* — Where a continuous variable is to be shown, solid black and white and graded cross-hatched areas constitute the most satisfactory graphic form. Care should be exercised to secure gradations of intensity in black and white, corresponding closely to the gradations of the variable.

5. *For showing time variations:*
   a — *Straight-line graph* — In general, the use of the straight-line graph between plotted points is to be recommended. . . .
   b — *Position of scales* — Intervals of time should be scaled invariably along the horizontal axis. . . .
   c — *Zero of vertical scale* — There is a strong presumption in favor of the appearance of the zero of vertical scale on the chart. . . .
   d — *Logarithmic scale* — The logarithmic scale vertically is to be used when rates of change or proportionate increases or decreases are to be emphasized. When the logarithmic scale is employed, the limits of the scale should be at some power of ten.

Statistical Standards in the Graphic Presentation of Facts [1]

The excuse for the use of graphics in statistical analysis is largely if not wholly their universal appeal. Graphs speak a common but frequently an inarticulate and confused language. There is an attractiveness about them which

[1] Adapted from Secrist, Horace, "Statistical Standards in Business Research," *Quarterly Publications of the American Statistical Association*, March, 1920, pp. 54–55.

is alluring but often deceptive. Their appeal is visual and instantaneous, not necessarily reasoned and reflective.

Distinguishing between rules for graphic presentation and the standards which give pertinency to the rules, the following standards may be formulated.

First. — A statistical fact and its form of representation should agree. By this single standard, deception, whether resulting from a confusion of the apparent with the real, or the superficial with the fundamental, is fully provided against. The object of statistical, like other analysis, is the establishment or determination of truth. Standards for graphics provide for their use in influencing but never in deceiving men. In spite of the standards adhered to, however, both results may be accomplished by the same graphic device.

Second. — Graphic forms should be selected according to their psychological appeal and their ease of comprehension, care always being taken not to violate the first standard.

Third. — Graphic forms should be chosen in accordance with (*a*) the form and complexity of the subject matter illustrated, and (*b*) the type of consumer for whom they are intended, or the purpose which they are intended to serve.

Fourth. — Graphic devices should be considered more as illustrations of analysis than methods by which analysis is made.

Fifth. — Graphic figures should be drawn as accurately as a visual representation will permit. Accuracy, of course, is never absolute. In graphics, the realization of relative accuracy of each part and of the totality is the standard set. To this standard for graphics, a corollary is needed; graphic forms should always be accompanied by the original data which they represent.

The Theory and Justification of Curve Smoothing [1]

The Theory of Smoothing Statistical Data. — It may often be known *a priori* that phenomena should exhibit a regular progression, and that data, when graphed, showing as zigzag lines, do not really represent the ideal fact, owing either to the paucity of the data, or to unavoidable error therein.

In a series of *group-values, i.e.* totals or aggregates between a series of limits of a variable, it is important to bear in mind that — assuming the counts on which they depend to be correct — what is *known* is merely the series of aggregates themselves; the probable distribution yielding these aggregates has to be *conjectured.* When the totals or aggregates are themselves regarded as subject to error, then the distribution may be modified within the limits of probable uncertainty, some groups being diminished and others, particularly adjoining ones, increased.

There are four principal classes of data to which the process of curve-smoothing is applicable. These may be indicated as follows:

(i) Frequencies of a phenomenon at successive epochs or during successive periods of time; as, for example, population estimates at given dates and numbers of deaths occurring during successive years.

(ii) Rates of occurrence of a phenomenon per unit of reference during successive periods; as, for example, birth-rates per thousand of population per annum for successive years.

(iii) Frequencies in respect of successive values of char-

[1] Adapted with permission from Knibbs, G. H., Commonwealth Statistician, "The Mathematical Theory of Population, of its Character and Fluctuations, and of the Factors which Influence Them, etc." *Appendix A*, Vol. I, *Census of the Commonwealth of Australia*, Melbourne, pp. 86–88.

acters capable of continuous variations; as, for example, the number of persons at each age recorded at a given census.

(iv) Rates of occurrence of a phenomenon per unit of reference in respect of successive values of characters susceptible of continuous variation; as, for example, rates of mortality per unit per annum during a given decennium in respect of each age.

In all these cases the characteristic of continuous variation is assumed to exist either actually or virtually. Where statistical results are *discontinuous* such a process is, strictly speaking, inapplicable; as, for example, in the tabulation of census population according to birthplace, occupation, or religion. In some cases, however, although the data are strictly speaking discontinuous, the principle may be applied partially; for example, in the case of a tabulation of dwellings according to number of rooms or according to number of inmates. In such cases the character possessed is *progressive without being continuous;* nevertheless, with proper qualifications, the smoothing principle may be applied even to these.

Another example, more nearly approaching but not attaining continuous variation, is the representation of dwellings according to rental value.

Object of Smoothing. — From the foregoing it will be seen that the data to which the smoothing process is strictly applicable are those which may be regarded as functions of a continuous variable. But whether such functions are readily expressible by means of algebraic formulæ or not, is, of course, really immaterial. The essence of the matter is that in any instance the data are in the main such as admit of representation by means of a continuous line, or a continuous surface or solid in relation to continuous units of reference. When such representation has been made of the crude results of observation, it is ordinarily found that the line surface or solid exhibits evidences of marked irregularities as between adjacent points or series of points, their general trend, however, suggesting an underlying basis of orderly progression. This progression is, of course, affected by minor influences operating at individual points, and is more or less masked by the paucity of the data on which the representation has been based; thus suggesting further that were it possible to obtain data of unlimited extent, these irregularities would become negligible. For this reason the object of the smoothing process may be said to be that of removing these apparently accidental irregularities, and of thus disclosing the basic or ideal uniformity which may be presumed to represent the facts in all their generality.

Justification for Smoothing Process. — The justifications for the smoothing process may thus be said to be:

(*a*) That the irregularity does not represent the phenomenon in its generality, since much of the observed irregularity is known *a priori* to be due only to paucity of data;

(*b*) or that it is known that the phenomenon subject to observation is really regular;

(*c*) or, again, that the observed data suggest that regularity of trend will not efficiently represent them.

It has been objected that any system of smoothing is, strictly speaking, unwarrantable, since such a process virtually attempts to make the facts accord with more or less questionable preconceptions regarding them. To this view it may be rejoined that if the process were such as to produce results which, though smooth, differed systematically and materially in their distribution from the original observations, the objection would be valid. Where, however, due consideration is given to the relative magnitudes of the original data, and the smoothed results accord therewith as closely as the data will allow when these exhibit a general trend, then the only preconception that can be regarded as operative is the justifiable one that ordinarily natural phenomena do not progress *per saltum*. In this connection it must be noted that where there is distinct evidence at any stage of a cataclysmic disturbance of results, the smoothing process for such points or periods will usually be invalid or not properly applicable. Examples of such cataclysmic disturbances of statistical data are war, famine, pestilence, earthquake, etc. Even in these cases, however, it appears admissible under certain circumstances to apply a smoothing process; as, for example, in cases where the disturbances referred to are of more or less frequent occurrence, and are not merely isolated instances.

One of the most cogent justifications for the smoothing process has its warrant in the fact that the recorded results of any statistical observations are necessarily approximative, and hence that the value of the function *recorded* for any given value of the variable is probably not usually more accurate than an estimate based on the recorded values in respect of preceding and succeeding values of the variable. This consideration suggests the idea of weighting successive observations to obtain most probable values, which idea forms the basis of one of the leading methods of adjustment. Again, where the results of the observations are to be employed as guides to future action, it is clear that these results should, as far as practicable, be freed from all fluctuations which may be considered merely accidental, and thus unlikely to be reproduced in future experience. This is of considerable importance in connection with the construction of mortality and sickness, superannuation, and similar tables to be used in the computation of rates of premium, and for the conduct of valuations.

REVIEW

1. How, if at all, does the above discussion apply to frequency series showing:

(1) The grades assigned to civil service applicants as a result of a written examination?

(2) The marks assigned as a result of an oral "mental test"?

(3) The number of workmen working classified hours?

(4) The number of brick two-story houses per unit of area in a residential district of city X?

2. How, if at all, does the above discussion apply to historical series showing:

(1) The number of troops embarking daily for France from the port of New York, June 1, 1918, to November 11, 1918?

(2) The daily total stock sales on the New York Stock Exchange, August 1, 1914, to October 1, 1914?

(3) The number of personal injuries in factory X from January 1, 1920, to June 30, 1920?

Some Advantages of the Logarithmic Scale in Statistical Diagrams [1]

The graphic method in statistics is primarily a device for presenting vividly the significant relations of phenomena. Each slope of a curve in an ordinary two-dimension statistical diagram is the visible expression of some relationship. If the purpose of a particular statistical presentation is simply an accurate recording of separate details, a diagram is, of course, a poor substitute for plain numerical statements; but when the relative aspects of the data are to be emphasized the diagram comes into its own.

[1] Adapted with permission from Field, J. A., "Some Advantages of the Logarithmic Scale in Statistical Diagrams," *Journal of Political Economy*, October, 1917, pp. 806–841.

And yet, even within this sphere of its special excellence, graphic representation, in terms of the common, natural scale of uniform intervals, has very real limitations. Too frequently, though the problem is simple and the diagram is well done, the eye will fail to detect the precise nature of the relationship which the statistician seeks to present.

Some of the shortcomings of natural-scale representation are fairly illustrated by Diagram I. The upper and

DIAGRAM I. — NET DEPOSITS (HEAVY LINE) AND RESERVES (LIGHT LINE) OF THE CLEARING-HOUSE BANKS OF NEW YORK CITY, ACCORDING TO THE 41ST WEEKLY REPORT (EARLY OCTOBER) IN EACH YEAR, 1867–1909

Natural Scale

Data (except for the year 1888) from *Statistics for the United States, 1867–1909*, compiled for the National Monetary Commission by A. Piatt Andrew

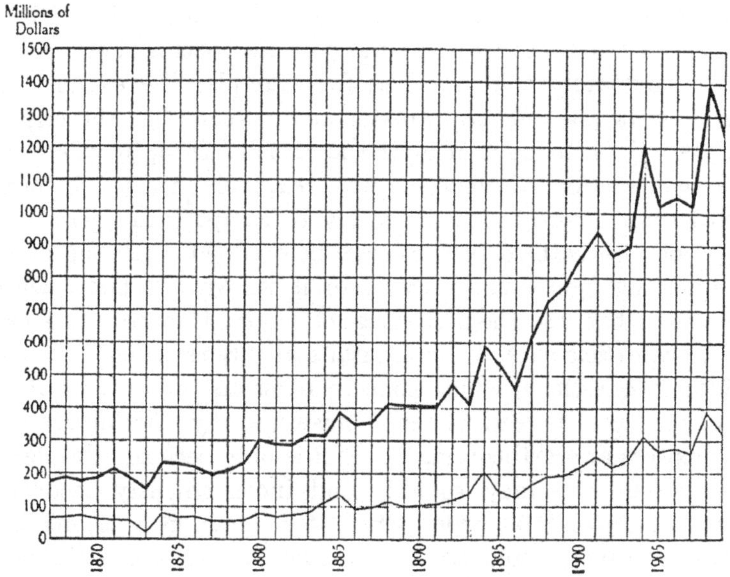

lower curves of this figure show, respectively, the net deposits and the reserves of the New York Clearing-House banks in early October of each year from 1867 to 1909, inclusive. From the diagram in this form certain facts are indeed sufficiently clear. Both deposits and reserves increased markedly during the period under review. The increase of each, though on the whole progressive, has been subject to appreciable fluctuations; and the fluctuations of one curve are associated with synchronous and apparently similar fluctuations of the other. The amount of deposits or of reserve in the early days of any particular October may be estimated by consulting the scale at the side of the diagram. The amount of increase or decrease of either item during a given year or term of years is not difficult to determine approximately. All this information, then, the ordinary scale gives adequately. Some of it would be less satisfactorily given by any other scale. But if we press our inquiries further and ask, on the basis of these early October statements, whether, for example, the expansion of deposits was *relatively* greater in the year after the crisis of 1907 than in the year after the crisis of 1873, or whether the contraction of deposits was relatively greater before 1873 than before 1896; if we try to compare the percentages of reserve held in the years before 1870 with the corresponding figures since 1895; or if we wish to know specifically what was the percentage of reserve in early October of 1905, deficiencies of the natural scale are revealed. None of these questions, which concern relations rather than detached facts, is satisfactorily answered by the diagram. If answers are forthcoming at all, it is only because, through the scales, one may roughly and inconveniently recover the numerical data from which the diagram was made. This, however, could have been more easily accomplished by ignoring the diagram altogether and consulting its data in the form of a table.

It is practicable, of course, to contrive a diagram, drawn to a natural scale, with the special purpose of bringing out some one fact or relation which in Diagram I has remained obscure. Thus the percentage of reserve of the New York banks could be plotted, year after year, as a separate curve. This curve, however, would in turn fail to show the absolute amounts of reserves and deposits. The difficulty is to devise a form of representation which shall show, directly and graphically, both relative and absolute magnitudes. A complete solution of this problem is hardly attainable, but logarithmic diagrams in certain cases go far toward meeting the want where the relative aspects of the phenomena are primarily to be emphasized.

The logarithmic scale may indeed be described as a scale of ratios. On it absolute distances measure relative magnitudes. The numbers which occur at equal intervals along a logarithmic scale thus form not an arithmetic but a geometric progression; and consequently the same proportionate relation exists between any two numbers a given distance apart on a given logarithmic scale, regardless of their absolute magnitudes and regardless of their absolute differences. Conversely, the numbers 2 and 4 on a logarithmic scale are separated by the same distance as the numbers 500,000 and 1,000,000, for the simple and decisive reason that the larger number of each pair is double the smaller number.

The mathematical principle of the scale is suggested by Diagram II. Here the graduations above the horizontal line mark off the intervals of a logarithmic scale from 1 to 100. The feature of this scale which at once strikes the eye rather bewilderingly is that the interval between successive numbers is not constant, but progressively narrows as the numbers grow larger. Closer scrutiny reveals the more significant and clarifying fact that the interval is constant between numbers which bear to each other a given ratio. Thus 1, 2, 4, 8, 16, 32 stand at equal distances apart; as do 1, 3, 9, 27, or 1, 5, 25, or 1, 10, 100. The uniform interval which separates the numbers of this last-named series — successive powers of 10 — has been taken as the

Diagram II. — The Logarithmic Scale from 1 to 100

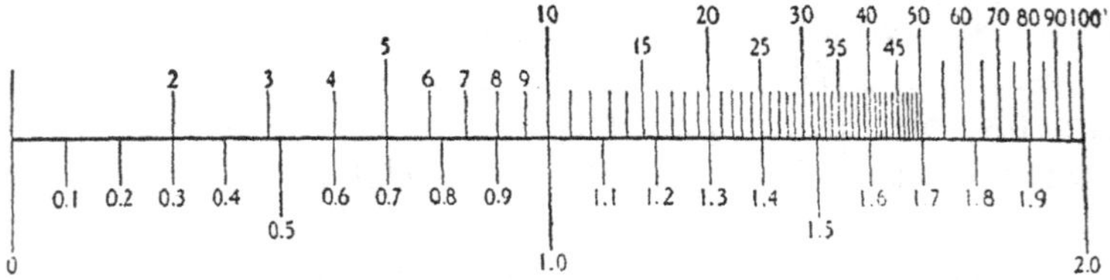

unit upon which is based the ordinary scale below the horizontal in the diagram. If, now, any number on the upper scale be regarded as a power of 10, it will be found that the corresponding reading of the lower scale gives the index of that power. This relation holds invariably; for not only do we find 10 (*i.e.* 10^1) opposite 1, 100 (*i.e.* 10^2) opposite 2, and 1 (*i.e.* 10^0) opposite 0, but the square root of 10 (*i.e.* $10^{\frac{1}{2}}$, or 3.1623) is opposite 0.5; the square root of 1000 (*i.e.* $10^{\frac{3}{2}}$, or 31.623) is opposite 1.5 — and so on indefinitely, whatever the index of the power. In fact, the number at any point of the lower scale is the common logarithm of the number at the same point of the upper scale.[1]

[1] The system of logarithms which is in ordinary use expresses any given number as a certain power of 10. The logarithm of the given number indicates what power of 10 that number is. Thus the logarithm of 10 is 1; the logarithm of 100, *i.e.* of 10×10, or 10^2, is 2; the logarithm of 1000, or 10^3 is 3, and so on. A logarithm is in fact an exponent — the index of a power — and the derivation and uses of logarithms consequently follow

If, now, it is desired to use the logarithmic scale in the construction of a statistical diagram, we may proceed in either of two ways. We may reduce the data to logarithmic terms, and then, using an ordinary natural scale, plot the logarithms of the given quantities instead of the quantities themselves. Or if we have at our disposal coordinate paper ruled at logarithmic intervals, like the intervals of the upper scale of Diagram II, we may work directly, without any reduction of the data, locating the points of the diagram quite mechanically by the graduations of the paper, and relying upon these graduations for the logarithmic character of the result. The two methods are entirely equivalent, as should be evident from Diagram II. Indeed it is often convenient to regard a diagram as constructed by both methods, and to supply for its more complete explanation a logarithmic scale of the natural num-

the algebraic rules of exponents. In the case of a number which is not an even power of 10 it is possible to compute the logarithm in the form of a fractional exponent. For example, as the text implies, the logarithm of 31.623, the square root of 1000, *i.e.* $\sqrt{10^3}$ or $10^{\frac{3}{2}}$, is 1.50. By extending the principle of fractional exponents the logarithm of any assignable number may be approximately expressed.

The peculiar advantage of the logarithmic scale in statistical work is a consequence of the elementary logarithmic principle that the difference between the logarithms of two numbers is the logarithm of the ratio of the one number to the other. That is,

$$\log a - \log b = \log \frac{a}{b}.$$

Hence, whenever the ratio between two numbers, a and b, is the same as the ratio between two other numbers, p and q, so that $\frac{a}{b} = \frac{p}{q}$, and $\log \frac{a}{b} = \log \frac{p}{q}$, it will follow that $\log a - \log b = \log p - \log q$. Plotted to a given natural scale, $\log a$ and $\log b$ would thus differ by the same interval as $\log p$ and $\log q$, — the equality of these differences indicating the equality of the ratios $\frac{a}{b}$ and $\frac{p}{q}$. The device of plotting statistical quantities in terms of their logarithms is, then, simply an exploiting of the general principle that the absolute difference between two logarithms is a measure of the relative difference of the numbers to which they correspond.

bers on one side, and a natural scale of their logarithms on the other.[1]

Before attempting a logarithmic presentation of the bank data of Diagram I, it will be well to consider, in artificially simplified cases, certain general properties of logarithmic diagrams which furnish the key to their interpretation.

Diagram III. — Arbitrary Example of a Phenomenon Increasing by Equal Relative Oscillations

Natural Scale

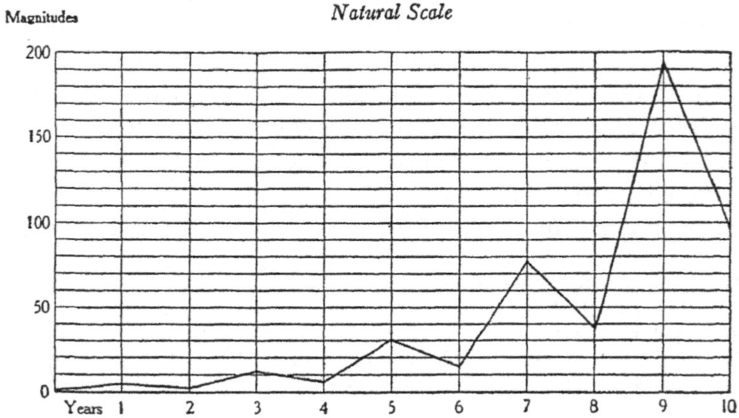

Let us take for our first illustration the arbitrary example of Diagram III. Here an assumed phenomenon, which has a magnitude of 1 when it is first observed, increases to 5 in the course of a year and then, in the second year, falls off to $2\frac{1}{2}$. In the third year it again increases fivefold, to $12\frac{1}{2}$. In the fourth year it again declines by half, to $6\frac{1}{4}$. Thus alternately quintupled and cut in two, the phenomenon grows by perfectly regular oscillations. Diagram III, which is drawn to an ordinary natural scale, shows vividly the accelerated character of this increase, stated in abso-

[1] For an example of this treatment see Diagram IV on p. 289.

lute numbers; but precisely because it is a natural-scale diagram it fails to show at all obviously that the rate of relative rise and fall is the same for all the oscillations. The earlier waves of the curve, which are absolutely small, are made to seem in all respects comparatively insignificant.

DIAGRAM IV.—ARBITRARY EXAMPLE OF A PHENOMENON INCREASING BY EQUAL RELATIVE OSCILLATIONS

Logarithmic Vertical Scale

Data of Diagram III

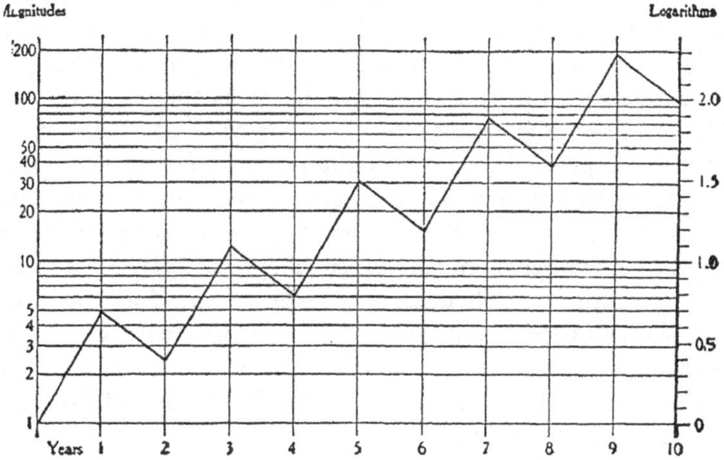

Strikingly different is the effect of Diagram IV, in which the data of Diagram III are plotted to a logarithmic scale. Absolute magnitudes here can be determined only from the numbers of the scale: the graphic evidence of the diagram establishes the identity of the relative changes, step by step, for the whole serrate curve. Every ascent has the same vertical rise. That is, the indicated percentages of increase are uniform. Each decline has the same drop: the percentage of decrease shown by each is the same.

This equal relative significance of equal absolute distances is the essential characteristic of the logarithmic scale.

Certain fairly obvious but important corollaries follow from this fundamental principle. Since the upstrokes of the curve in Diagram IV are all straight lines rising by the same amount, and since each rise, occurring in the same period of time, is allotted in the diagram the same horizontal distance, it follows that the slope of the several upstrokes is the same. The downstrokes are similarly all of the same slope. Quite generally, where a curve is drawn to a logarithmic vertical scale and a natural horizontal scale, equal slopes indicate equal rates of relative change. By extension of this rule it will be seen that a constant rate of increase is represented in a logarithmic curve by a constant slope — *i.e.* by a straight line; and that wherever in such a logarithmic diagram two curves run parallel, in the sense that the vertical distance between them remains unaltered, the phenomena which they respectively represent maintain to each other a constant ratio, inasmuch as any change of the one is evidently coincident with a change of the other to the same relative extent.

These generalizations may be simply illustrated by the examples which follow.

In Diagram V, drawn to natural scale, the continuous curve traces the growth of the population of the United States, according to the decennial enumerations of the United States Census from 1790 to 1910, inclusive. The broken line, uppermost in the diagram, shows what the growth of population would have been if the rate of relative increase observed between 1790 and 1800 — 35.1 per cent for the decade — had persisted without change since that time. The dotted line at the bottom of the figure shows what the growth would have been if the absolute increase

DIAGRAM V. — GROWTH OF THE POPULATION OF THE UNITED STATES, 1790–1910

The continuous line shows the actual increase according to the census returns. The broken and dotted lines show the growth which would have taken place if relative and absolute increase, respectively, had continued at the rate of the first decade.

Natural Scale

Data from 13th Census of the United States, I, 24. The corrected estimate for 1870 has been taken instead of the original enumeration.

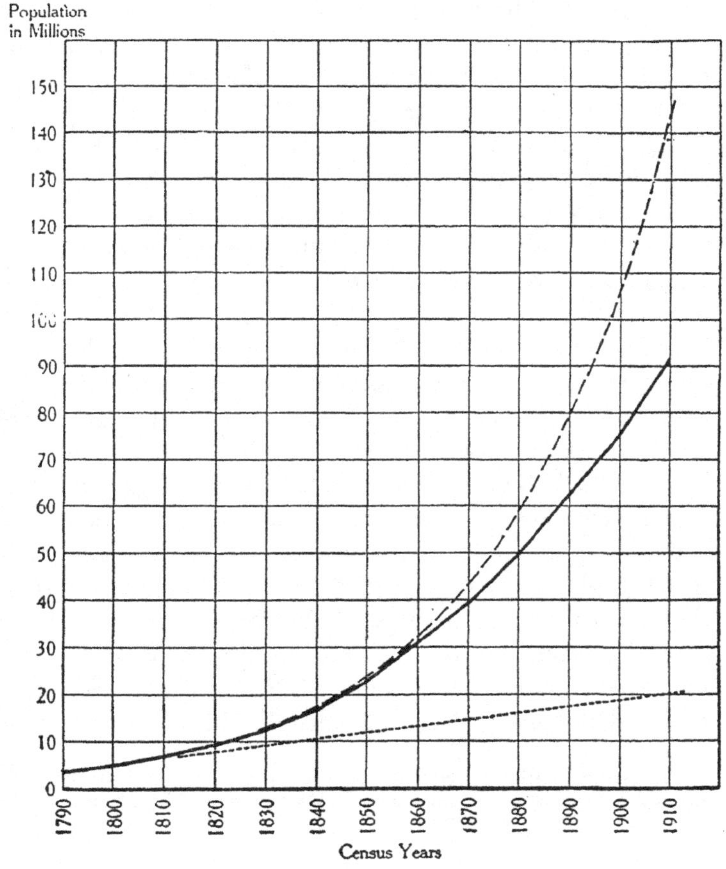

of population in each decade since 1800 had been the same as the increase — 1,379,269 persons — from 1790 to 1800. In other words, these two additional curves represent respectively geometric and arithmetic progressions based on the observed increase in the first intercensal period. It is to be noted that in a natural-scale construction the curve of arithmetic progression is a straight line.

In Diagram VI, drawn to a logarithmic scale, the continuous line, the broken line, and the dotted line represent each the same data as in Diagram V. But here the character of the curves is significantly different. The dotted arithmetic-progression curve, recording a constantly diminishing ratio of increase, falls away in this figure more and more toward the horizontal. And here it is the geometric progression which appears as a straight line, its constant slope denoting a constant rate of increase — *i.e.* the same relative increase in every equal period of time.

The growth of funds invested at compound interest affords another instance of geometric increase and therefore another example of a straight-line curve if a diagram is drawn to a logarithmic scale. The slope of the curve here depends upon the rate of interest and the interval between dates at which the interest is regularly compounded; but for a given rate and interval it is fixed and constant. Hence, a logarithmic chart equivalent to a compound-interest table may very readily be constructed. Diagram VII is such a chart. In it a single straight line suffices to indicate the amount to which an initial sum of $100, compounded semi-annually at a given rate, will have increased on any compounding date included in the diagram.[1] The

[1] The period of time covered by such a chart is of course in principle unlimited, for the lines will continue with their same specific slopes however far the diagram may be extended.

DIAGRAM VI. — GROWTH OF THE POPULATION OF THE UNITED STATES, 1790–1910

Logarithmic Vertical Scale

Data and explanations as in Diagram V

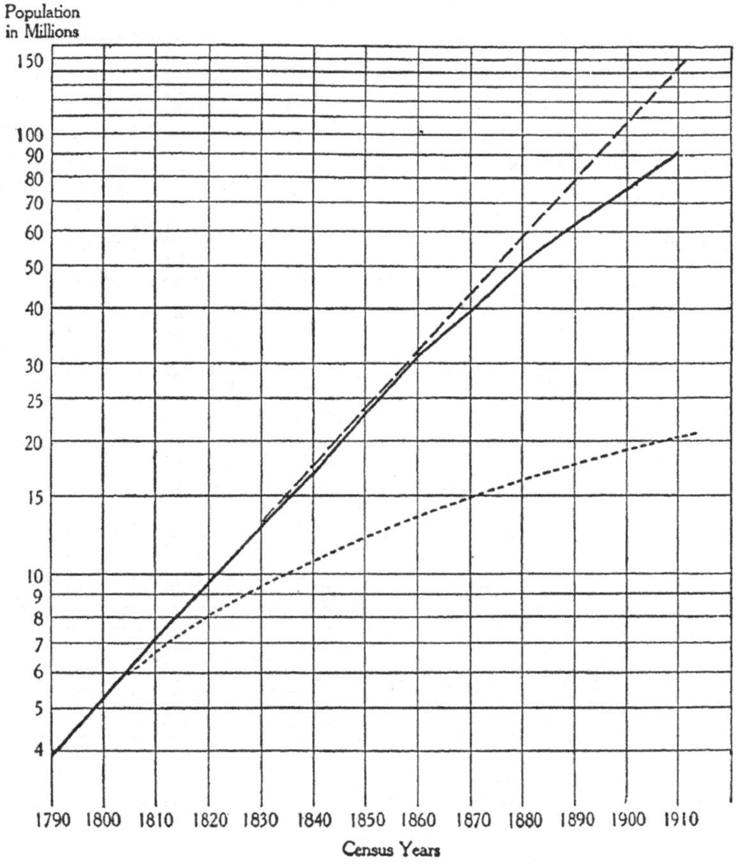

DIAGRAM VII. — COMPOUND-INTEREST CHART (SEMIANNUAL COMPOUNDING)

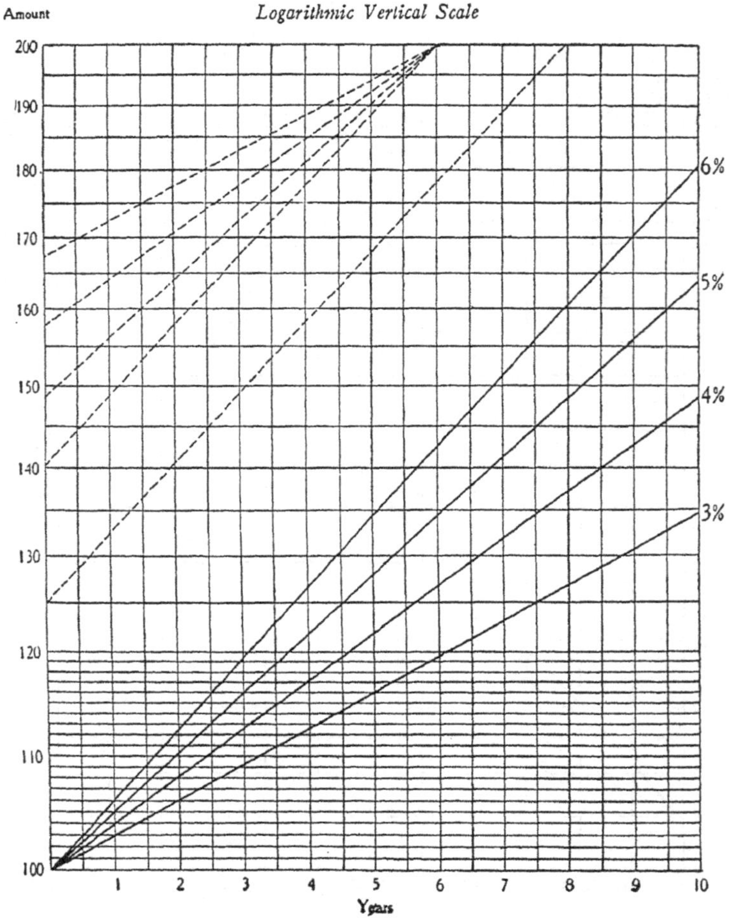

4 per cent line is steeper than the 3 per cent line; the 5 and 6 per cent lines are successively steeper still; but all are straight, and for each, when the scales of the diagram are once determined, the slope is fixed and characteristic.

The same diagram serves also to illustrate another property of logarithmic diagrams that has already been mentioned. The broken line across the middle of the figure has been drawn to show the increase of $125, compounded semiannually at 6 per cent. It is at once apparent that this line parallels the continuous line of the increase of $100 at the same rate. The reason for the parallelism is tolerably patent. Each of the sums, $100 and $125, increases every six months by 3 per cent of its accumulated amount. That is, each sum is semiannually multiplied by 1.03. In the diagram, therefore, each of the two lines must rise, from one ordinate to the next, by the fixed vertical distance which, on the logarithmic scale, corresponds to the ratio 1.03 : 1.00. This, of course, insures that both rise alike. Or it may rather be argued that since original sums in the proportion of 1.25 to 1 are here assumed to be compounded at the same rate and the same interval, the cumulative results will be at any subsequent time in the same proportion of 1.25 to 1. The vertical distance between the two curves on any ordinate must therefore express the ratio 1.25 : 1.00, and hence, since a given ratio always corresponds to the same absolute interval on a logarithmic scale, the curves must be always at the same distance apart and therefore parallel. It follows that if a point be taken on the initial ordinate of this diagram, opposite the value $125 of the vertical scale, the straight line drawn through that point parallel to the original 6 per cent curve will represent the compound increase of $125 at 6 per cent. Similarly, to find the increase of any capital sum at any rate of compound interest, one has only to draw a straight line starting at the height which denotes the given sum and running parallel to a standard curve for the given rate of interest. In Diagram VII this principle has a somewhat different application. Through the point representing a sum of $200 at the end of 6 years have been drawn broken lines parallel to the standard curves showing respectively 3 per cent, 4 per cent, 5 per cent, and 6 per cent increase. These several broken lines cut the initial ordinate at heights which, read in terms of the vertical scale, show what amount of money, compounded semiannually at each respective rate of interest, would amount to $200 after 6 years. . . .

Since logarithmic scales have no zero, logarithmic diagrams can have no base-line at zero. Indeed, they have no base-line at all; or, rather, every value of the logarithmic scale is as much a base-value as any other. This follows from the cardinal principle already repeatedly stated, that the same absolute interval stands for the same ratio of magnitudes at any and every part of a given logarithmic scale. It obviously constitutes an essential distinction between logarithmic and natural-scale diagrams. In a natural-scale diagram the importance of showing the base-line at zero of the vertical scale can hardly be urged too strongly. If this base-line be omitted, as it often is in unintelligent work, proper visual estimation of relative magnitudes is made impossible. Such omissions in complex natural-scale diagrams involving more than one base-line lead to extreme confusion and fallacy. In logarithmic diagrams fallacious effects of this particular sort are impossible; but any suggestion of a specific base-line may prove disconcerting to those unfamiliar with the logarithmic scale and may cause misconception of its character.

The principles which have thus far been developed may now be recapitulated:

Throughout a given diagram, and regardless of the absolute magnitudes concerned:

(1) a given distance between any two points, measured along a logarithmic scale, indicates in every case the same ratio between the two magnitudes which the positions of the points represent;

(2) when changing magnitudes are plotted to a vertical logarithmic scale, and unit intervals of time are plotted to a horizontal natural scale,

(*a*) the slope of a curve is always an index of the rate of relative change;

(*b*) a straight line represents a constant rate of relative change; and, conversely, a constant rate of relative change is always represented by a straight line;

(*c*) where the vertical distance between two curves is constant the variables which they respectively represent maintain always the same proportion one to the other; and, conversely, two variables constantly in the same proportion are always represented by two curves at a fixed vertical interval.

The logarithmic scale admits of no zero, and in terms of a logarithmic scale no base-line should ordinarily be indicated.

With these general principles in mind we may now consider Diagram VIII, in which the bank statistics of Diagram I are plotted to a logarithmic scale. The questions which Diagram I failed to answer find here a ready solution, and incidentally illustrate certain useful devices for the interpretation of logarithmic diagrams in general.

The relative expansion of deposits, evidenced by the absolute rise of the upper curve in Diagram VIII, was plainly greater in the year following October, 1873, than in the year following October, 1907. How great it was in either year may be determined with the aid of the percentage scale of increase at the right of the main figure. This scale, it is to be noted, holds good for vertical measurements at all parts of the diagram, since its logarithmic intervals make it a scale of ratios, quite independent of absolute magnitudes. The vertical rise of the deposit curve following 1873 shows by the scale an increase of approximately 50 per cent. The rise after 1907, similarly measured, is some 38 per cent.

Relative decreases of deposits can be tested in a manner quite analogous by the logarithmic scale of percentage decrease. Here, for convenience, the scale reads from the top downward, rather than up from the bottom, as in the scale of increase. The contraction of deposits from October, 1871, to October, 1873, as measured by the decrease scale, was about 27 per cent — appreciably greater than the contraction of some 22 per cent during the two years preceding October, 1896.

The proportion of reserve to deposits at any given date is obviously to be determined from Diagram VIII by measuring the appropriate vertical distance between the reserve curve and the deposit curve. For this purpose one might use the scales designed to measure increase and decrease. Thus, in October, 1905, deposits were not quite four times as great as reserves, according to the multiple scale. Interpreted by the scales of decrease, reserves were equivalent to slightly more than a quarter of the deposits, or were some 74 per cent less than the deposits. None of these statements, however, expresses reserves in the conven-

Logarithmic Vertical Scale

Data of Diagram I

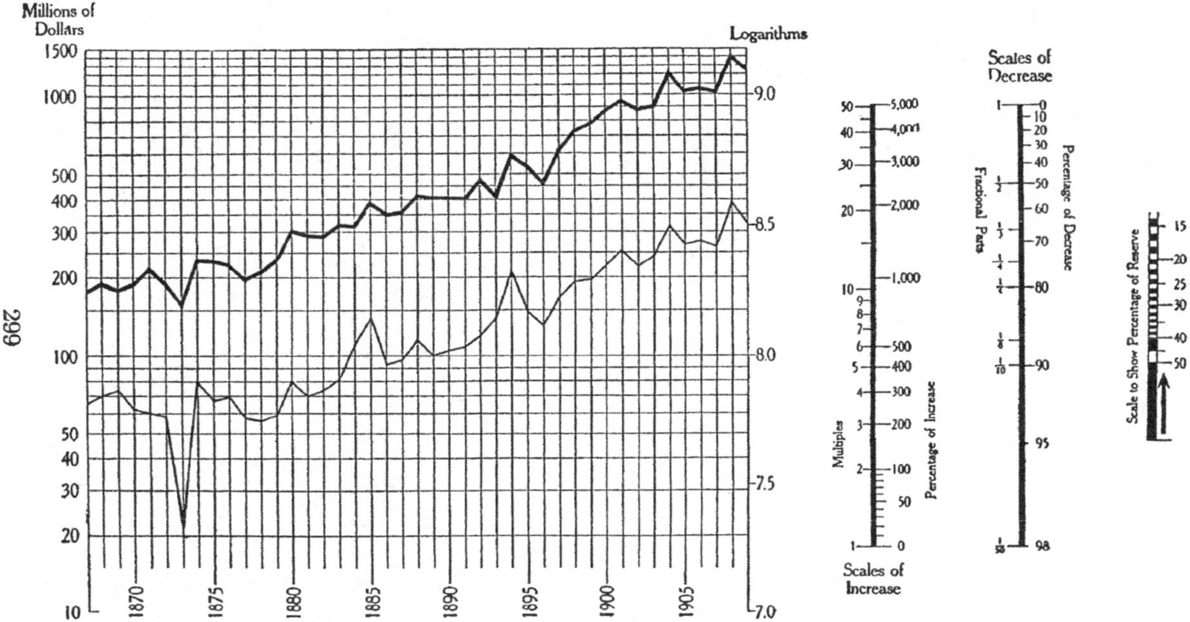

DIAGRAM VIII.—NET DEPOSITS (HEAVY LINE) AND RESERVES (LIGHT LINE) OF THE CLEARING-HOUSE BANKS OF NEW YORK CITY, ACCORDING TO THE 41ST WEEKLY REPORT (EARLY OCTOBER) IN EACH YEAR, 1867–1909

tional way as a percentage of deposits. For convenience, therefore, a special inverse logarithmic scale is provided at the extreme right of the figure. If a given vertical interval between the reserve curve and the deposit curve is laid off on this scale, from the bottom upward, the reading of the inverse scale states the reserve directly as a percentage of deposits. In October, 1905, it thus appears that the reserve stood at 26 per cent. The rough parallelism of the two curves throughout their whole course shows that the percentage of reserve has not greatly changed. Nevertheless, it is tolerably clear that the reserves held in early October were rather larger before 1870 than since 1895; for in the former period the curves are nearer together. The last of the questions which Diagram I left unsettled thus finds its answer in Diagram VIII. . . .

Another merit of no slight importance is to be recorded for the logarithmic scale: it is far superior to the natural scale for effecting comparisons when very small and very large quantities must be taken into account concurrently. . . . Whenever a historical curve records extreme growth, the same advantage is found. It is not necessary to dwarf the small beginnings in order to keep the later development within manageable dimensions. A study of Diagrams III and IV will illustrate this point. More striking illustration is offered in Diagrams IX and X. The production of tinplate in 1891 and the years immediately following was so small that the ordinary diagram (Diagram IX) leaves inconspicuous the extremely rapid rate of progress in output during those first years. The logarithmic diagram (Diagram X) quite reverses the emphasis. Plainly, the recent increase has been far from proportionate to the exuberant growth of the infant industry.

Although the years of small beginnings in a historical

record may present no features that require special consideration, the logarithmic historical diagram is again advantageous whenever substantially the same rate of relative increase characterizes the whole period under review. In

DIAGRAM IX. — ANNUAL PRODUCTION OF TINPLATE IN THE UNITED STATES, 1891–1912

Natural Scale

Data from D. E. Dunbar: *The Tin Plate Industry*, p. 15

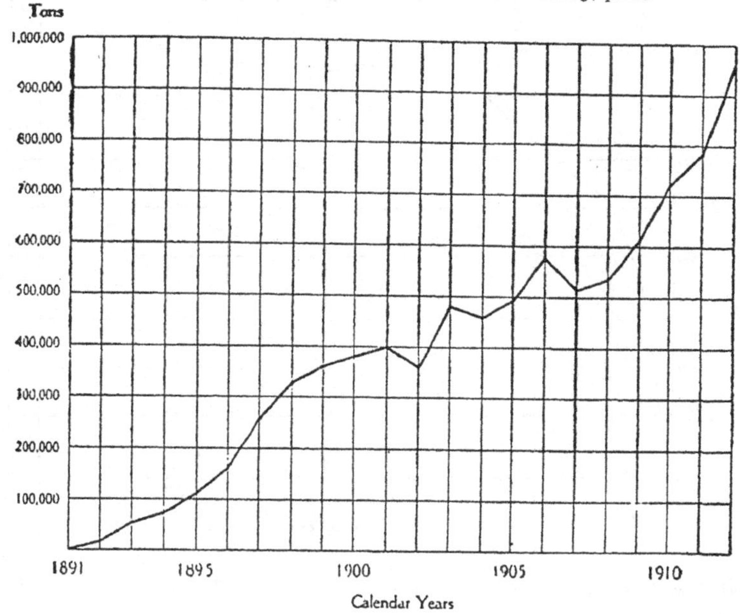

such cases the general trend or growth-axis of the logarithmic curve will of course be nearly straight. This is interesting for its evidence of consistent growth. It has the further technical merit of permitting the trend of the curve to be approximately maintained throughout at any desired slope by the mere choice of dimensions for the diagram. Hence such curves can readily be kept close to an inclination of 45°, with the result that irregularities of direction are much more easily noticed than if the slope were

DIAGRAM X.—ANNUAL PRODUCTION OF TINPLATE IN THE UNITED STATES, 1891–1912

Logarithmic Vertical Scale

Data of Diagram IX

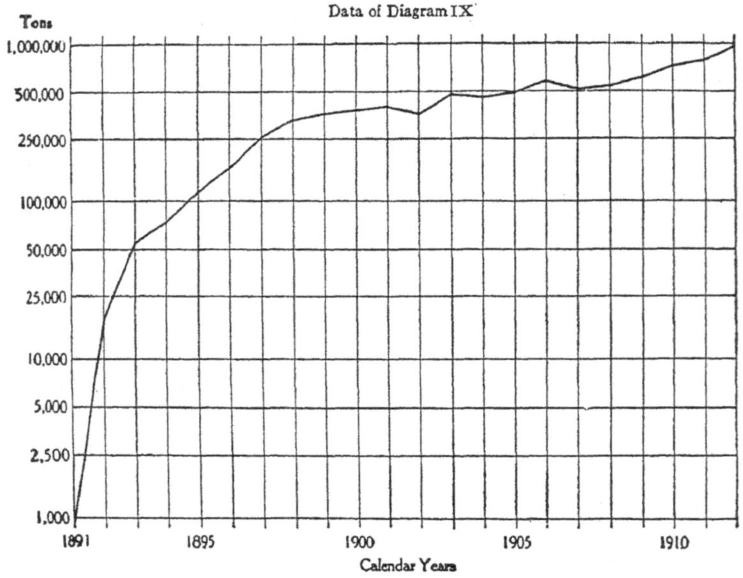

as steep or as flat as in natural-scale diagrams some parts of the curve often must be.

For the plotting of index-numbers logarithmic diagrams are particularly appropriate, for here the numbers themselves are ratios, and their relative aspect is important. If an index number of general prices should rise from 80 to 100, and later from 100 to 120, the two changes would appear of equal significance in an ordinary diagram. Yet the first is an increase of 25 per cent, the second, an increase of but 20 per cent. In their effects upon the purchasing power of stated money incomes the two changes are by no means the same. A logarithmic diagram reveals

DIAGRAM XI. — COURSE OF THE GENERAL INDEX NUMBER OF WHOLESALE PRICES PUBLISHED BY THE UNITED STATES BUREAU OF LABOR STATISTICS, 1890–1914

(AVERAGE PRICES FOR THE PERIOD 1890–99 ARE TAKEN AS 100)

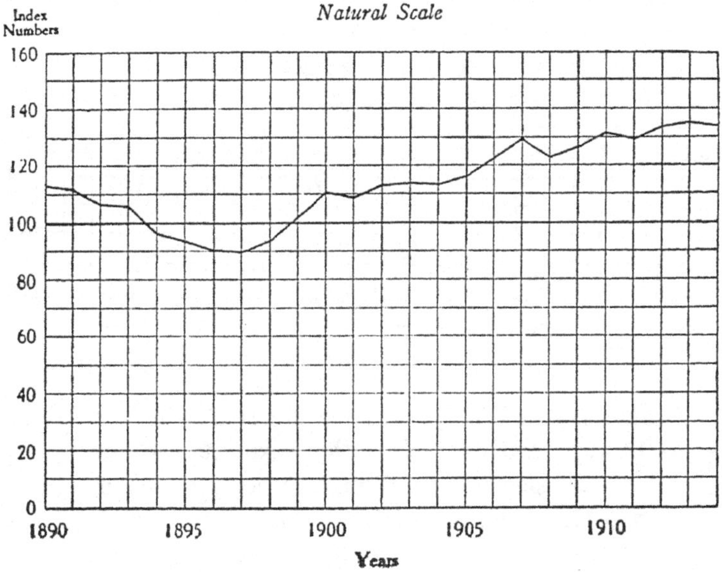

their significant difference. Diagrams XI and XII contrast the natural-scale method with the logarithmic-scale method in the case of the general index number of wholesale prices from 1890 to 1914, published by the United States Bureau of Labor Statistics. It will be remarked that the logarithmic figure, which does not require a zero base-line in order to convey a true sense of relative values, permits a considerable saving of space. . . .

From the illustrations which have been offered it will have appeared first of all that logarithmic diagrams present ratios and relative changes as directly and simply (though not, to the uninitiated eye, so obviously) as natural-scale diagrams present absolute differences. Consequently the

Diagram XII. — Course of the General Index Number of Wholesale Prices Published by the United States Bureau of Labor Statistics, 1890–1914

Logarithmic Vertical Scale

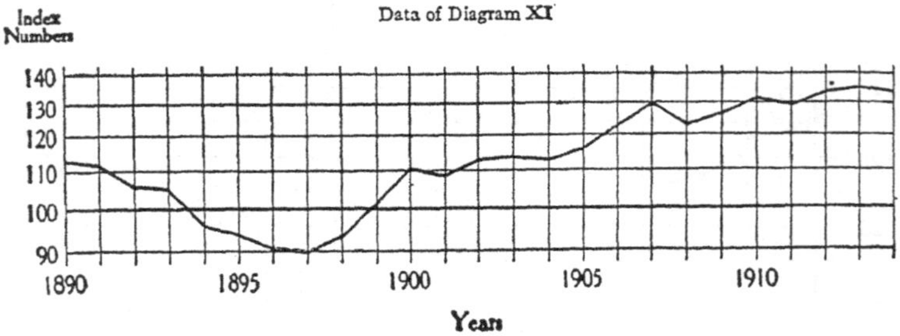

logarithmic method is peculiarly effective when the data are essentially relative; when they exhibit a tendency to increase or decrease at a fixed relative rate; or when significant proportionalities between different series of data are to be demonstrated. Incidentally it serves to economize space, and thus permits the inclusion of very diverse magnitudes in the same figure. These are real advantages, which clearly justify the use of logarithmic constructions in a considerable range of graphic work — sometimes by themselves, sometimes in conjunction with other forms of representation. How extensively such constructions will or should supplant ordinary figures on the natural scale need not now be argued. It is enough to make known their fundamental properties. When these are generally appreciated, we may trust the ingenuity and judgment of statisticians to find for logarithmic diagrams the place that they deserve.

REVIEW PROBLEMS

Diagrams and Graphs

1. From the data in the table below

Draw bar diagrams comparing the foreign holdings of common and preferred stocks for 1914 to 1919, inclusive.

Distribution of Foreign Holdings of the United States Steel Corporation Stock

Year	Foreign Holdings	
	Common	Preferred
Total	3,646,992	1,167,325
1919	368,895	138,566
1918	491,580	148,225
1917	484,190	140,077
1916	502,632	156,412
1915	696,631	274,588
1914	1,193,064	309,457

2. From the data in the table below

Draw two types of component-parts diagrams for holdings of common stock, showing for 1919 the proportion held by each country.

Foreign Holdings of Shares of United States Steel Corporation Common Stock, by Countries and by Years

Countries	Years						
	Total	1919	1918	1917	1916	1915	1914
Total . . .	3,680,002	358,912	480,163	476,675	496,516	687,177	1,180,559
Canada . . .	246,870	35,686	45,613	41,639	31,662	38,011	54,259
England . .	1,769,873	166,387	172,453	173,074	192,250	355,088	710,621
France . . .	237,424	28,607	29,700	30,059	34,328	50,193	64,537
Germany . .	6,932	959	891	612	628	1,178	2,664
Holland . .	1,398,655	124,558	229,285	229,185	234,365	238,617	342,645
Ireland . . .	5,833	160	19	19	914	1,730	2,991
Italy	1,548	281	281	281	279	280	146
Spain . . .	3,939	555	549	300	510	800	1,225
Sweden . . .	296	70	80	64	68	13	1
Switzerland .	8,632	1,649	1,292	1,442	1,512	1,267	1,470

3. Using the following data showing the number and deadweight tonnage of vessels employed in the West Indian and South American West Coast trades, graphically compare, by using *percentages*, frequency graphs of

(1) The number of vessels engaged.
(2) The tonnage of vessels engaged.
(3) Express the comparison in some other satisfactory form.

Table Showing the Distribution of Steam Freighters Classified by Size, Trading between the United States and the West Indies and with the West Coast of South America

Classified Dwt. Tonnage of Vessels	Vessels trading between the United States and			
	West Indies		South American West Coast	
	Number	Aggregate Dwt. Tons	Number	Aggregate Dwt. Tons
Total	99	281,900	75	434,941
500– 1,500	14	17,745	3	3,579
1,500– 2,500	40	79,593	5	10,367
2,500– 3,500	20	57,530	4	11,100
3,500– 4,500	11	42,674	8	31,300
4,500– 5,500	6	29,453	9	43,536
5,500– 6,500	6	35,755	17	102,995
6,500– 7,500			10	68,538
7,500– 8,500	1	7,850	11	87,148
8,500– 9,500			5	45,888
9,500–10,500			2	16,140
10,500–11,500	1	11,300	1	11,350

4. Using the following data showing the Immigrant Aliens admitted into the United States

(1) Construct a cumulative historigram on an "up to and including" basis for the period in question.

(2) Determine from the graph the number of immigrants that came into the United States during the first quarter of the period,

the first half of the period, the first three-quarters of the period. Similarly, determine the proportion of the entire time required to bring in one-quarter of the total number, one-half of the total number, three-quarters of the total number. Arrange these measures in the form of a statistical table, and briefly describe them.

Month	1914	1915	1916	1917	1918
January		15,481	17,293	24,745	6,256
February		13,873	24,740	19,238	7,388
March		19,263	27,586	15,512	6,510
April		24,532	30,560	20,523	9,541
May		26,069	31,021	10,497	15,217
June		22,598	30,764	11,085	14,247
July	60,377	21,504	25,035	9,367	7,780
August	37,706	21,949	29,975	10,047	7,862
September	29,143	24,513	36,398	9,228	9,997
October	30,416	25,450	37,056	9,284	11,771
November	26,298	24,545	34,437	6,446	8,499
December	20,944	18,901	30,902	6,987	

5. Using the following data

(1) Draw an ordinary historical chart comparing the foreign holdings of common and preferred shares of stock of the United States Steel Corporation.

In which has the decrease been more marked? Can this question be answered from this type of chart? Why?

(2) Draw a "ratio" chart using ordinary "ratio" paper.[1]

In which type of shares has the rate of decrease been most marked? Compare charts (1) and (2). Which type seems best suited to illustrate the change in holdings? Why?

[1] "Ratio" paper may be secured from the Educational Exhibition Company, 26 Custom House St., Providence, R. I.; Keuffel and Esser Co., 127 Fulton St., New York City; and the Standard Graph Co., 32 Union Square, New York City.

FOREIGN HOLDINGS OF SHARES OF U. S. STEEL CORPORATION

Common			Preferred		
Date	Shares	Per Cent	Date	Shares	Per Cent
Mar. 31, 1914	1,285,636	25.29	Mar. 31, 1914	312,311	8.67
June 30, 1914	1,274,247	25.07	June 30, 1914	312,832	8.68
Dec. 31, 1914	1,193,064	23.47	Dec. 31, 1914	309,457	8.59
Mar. 31, 1915	1,130,209	22.23	Mar. 31, 1915	308,005	8.55
June 30, 1915	957,587	18.84	June 30, 1915	303,070	8.41
Sept. 30, 1915	826,833	16.27	Sept. 30, 1915	297,691	8.26
Dec. 31, 1915	696,631	13.70	Dec. 31, 1915	274,588	7.62
Mar. 31, 1916	634,469	12.48	Mar. 31, 1916	262,091	7.27
Sept. 30, 1916	537,809	10.58	Sept. 30, 1916	171,096	4.75
Dec. 31, 1916	502,632	9.89	Dec. 31, 1916	156,412	4.34
Mar. 31, 1917	494,338	9.72	Mar. 31, 1917	151,757	4.21
June 30, 1917	481,342	9.45	June 30, 1917	142,226	3.94
Sept. 30, 1917	477,109	9.39	Sept. 30, 1917	140,039	3.59
Dec. 31, 1917	484,190	9.52	Dec. 31, 1917	140,077	3.88
Mar. 31, 1918	485,706	9.56	Mar. 31, 1918	140,198	3.90
June 30, 1918	491,464	9.66	June 30, 1918	149,032	4.13
Sept. 30, 1918	495,009	9.73	Sept. 30, 1918	147,845	4.10
Dec. 31, 1918	491,580	9.68	Dec. 31, 1918	148,225	4.11
Mar. 31, 1919	493,552	9.71	Mar. 31, 1919	149,832	4.16
June 30, 1919	465,434	9.15	June 30, 1919	146,478	4.07
Sept. 30, 1919	394,543	7.76	Sept. 30, 1919	143,840	3.99
Dec. 31, 1919	368,895	7.26	Dec. 31, 1919	138,566	3.84

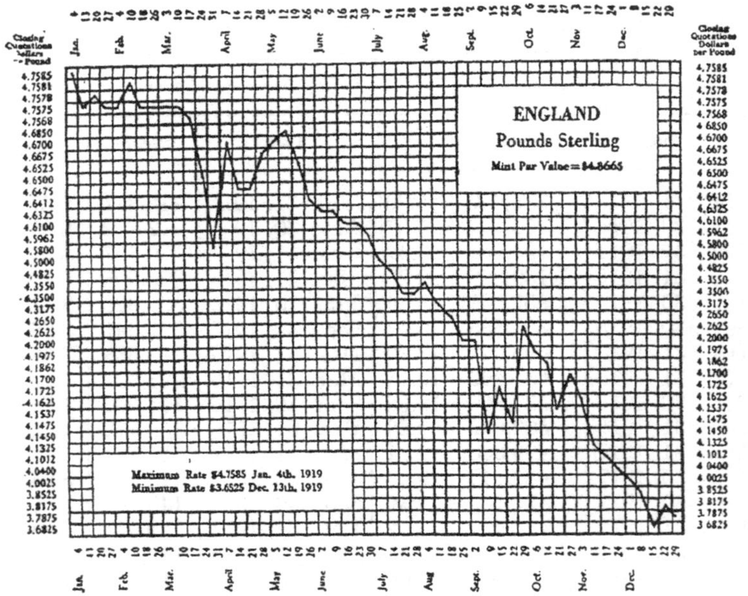

DIAGRAM I. — Purports to show the Trend of Sterling Exchange, 1919 [1]

6. Using the above chart

(1) Write a criticism of the method in which this curve is drawn.

(2) Redraw the curve according to the directions in the *Text* and the *Readings*. What differences do you note?

(3) Write a description of the trend of Sterling exchange based upon the original chart. How satisfactory is it? Why?

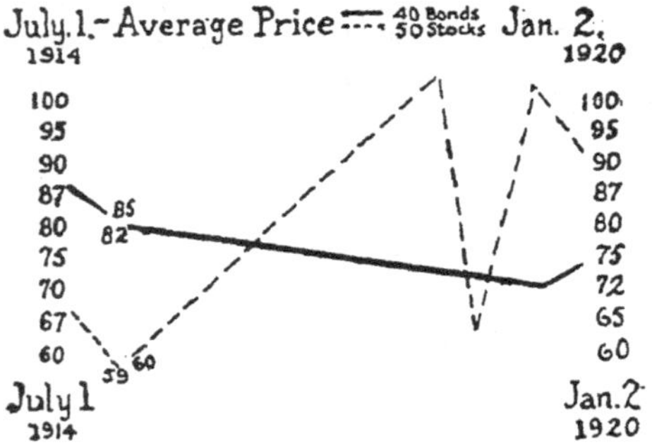

DIAGRAM II. — War Record of Bond Prices

7. Write a criticism of the above chart.

[1] Taken with permission from "Charts of the Fluctuations of Foreign Exchange Rates for the Year 1919," The First National Bank of Boston.

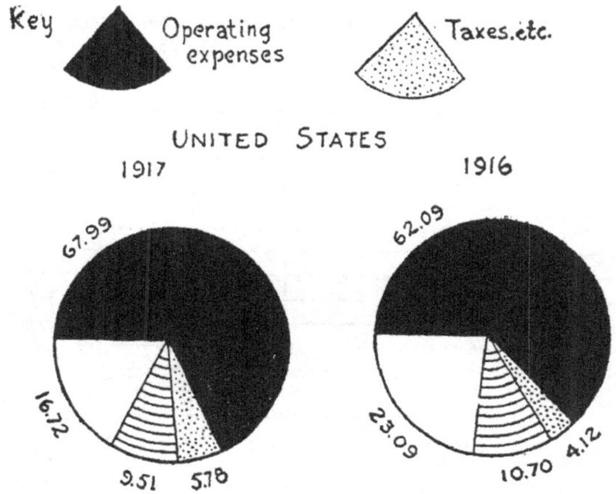

DIAGRAM III. -- Division of the Railway Dollar in October, 1917 and 1916

8. Using the above diagram

(1) Express the relationships by using some other form of component-part chart. How do you rank the relative methods? Why?

(2) Place the data in a table. In what ways, if at all, is the table a less satisfactory method of presenting the data?

(3) Express the relationships of the data in the form of a running statement of not more than 100 words.

DIAGRAM IV. — A Hundred Dollars' Worth of Cotton, 1913 to 1916

9. Using the above diagram

(1) Study the proportions of the figures. Are the figures drawn to scale? What is the scale?

(2) Redraw this diagram according to the rules discussed in the *Text* and *Readings*.

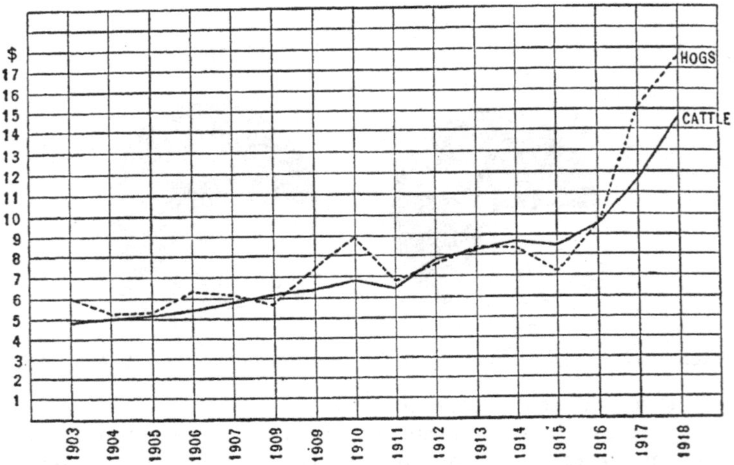

DIAGRAM V. — The Average Yearly Prices of Cattle and Hogs at Chicago, 1903 to 1918

10. Discuss the construction of the above diagram in the light of the discussion in the *Text* and *Readings*.

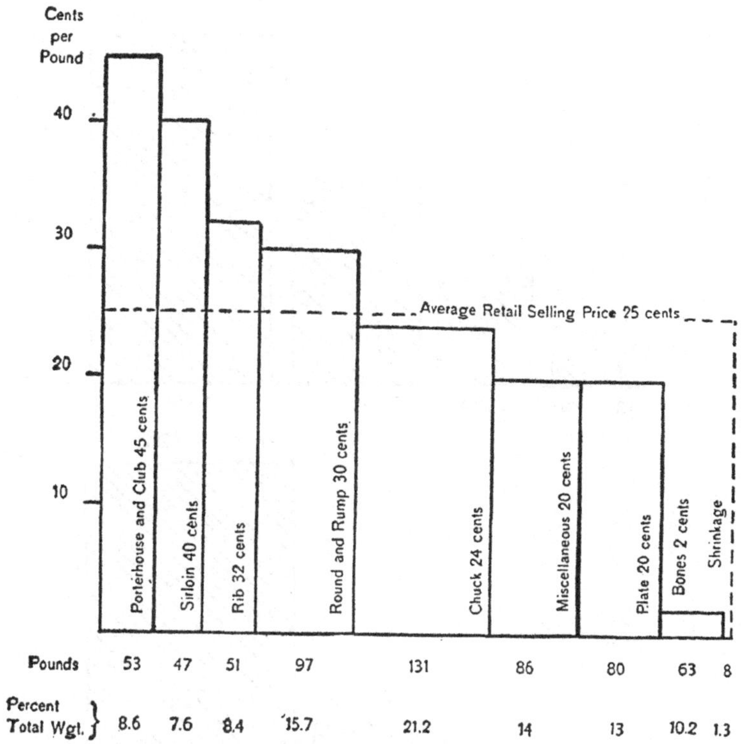

Diagram VI. — Weights and Retail Prices of the Different Cuts of Meat

11. Using the above diagram

(1) Describe the principles on which it is drawn.

(2) Write a paragraph descriptive of its contents.

(3) Put in the form of a table the data shown in the diagram.

(4) Which is the most effective, the description, the diagram, or the table? Why?

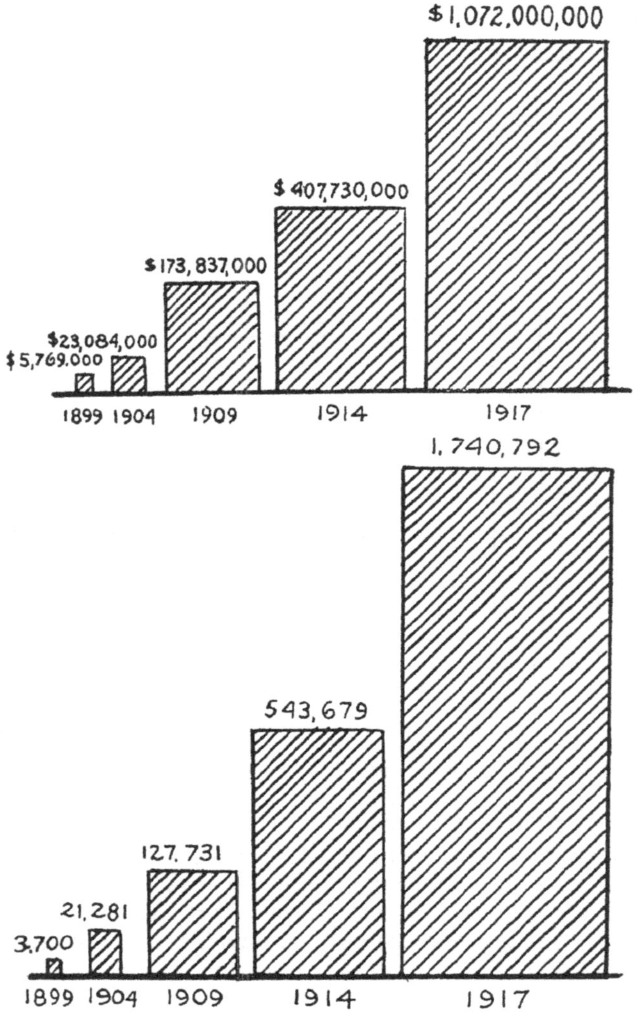

DIAGRAM VII. — Growth of the Automobile Industry and the Investment behind it, 1899 to 1917

12. Criticize the form of these diagrams, according to the standards established in the *Text* and *Readings*.

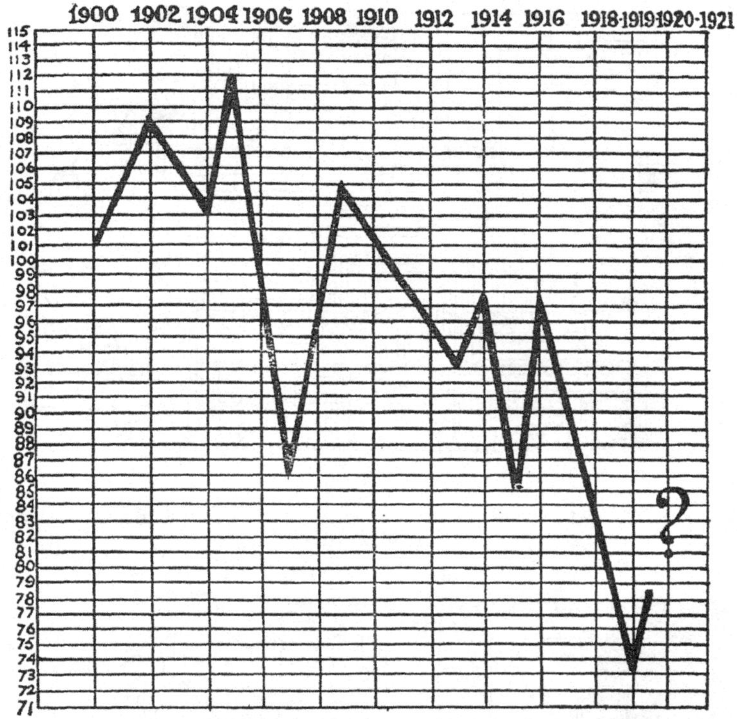

Diagram VIII. — Course of Average Price of 15 Standard Long Term Railroad Bonds During Past Twenty Years

13. Using the above diagram

(1) By what standards would you test its merits?

(2) Write a description of the trend of Railroad Bonds based upon this chart. How satisfactory is it? Why?

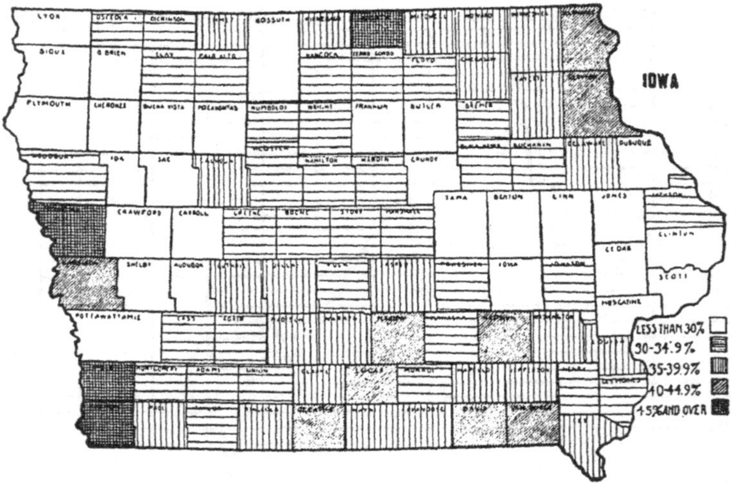

DIAGRAM IX. — Location of Share-rented Farms which include Stock-share Rented Farms

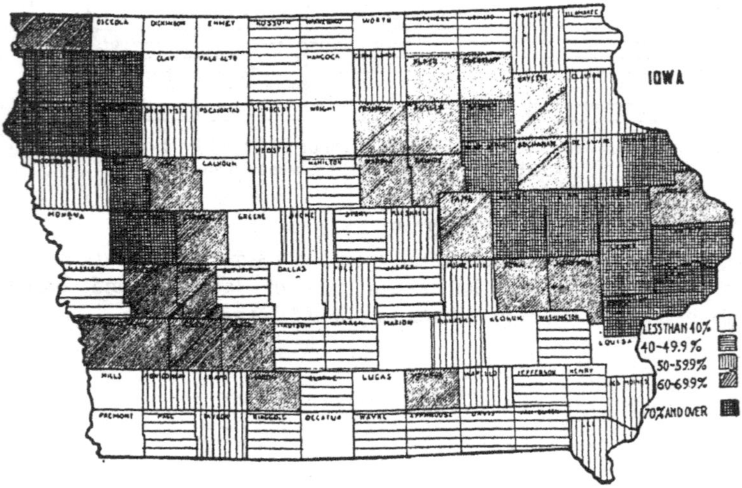

DIAGRAM X. — Location of Cash-rented Farms

14. Using diagrams IX and X,

(1) Criticize the methods by which they are drawn.

(2) In what way, if at all, would you criticize the titles? Be specific.

(3) Extract, from the maps, the concrete data, place them in a statistical table, and compare the data. How does the tabular method of presentation compare with the graphic?

(4) Secure, if possible, county outline maps of Iowa, and redraw the illustrations according to the instructions in the *Text*.

(5) Would it be possible or advantageous to use, in this case, any one of the type of dot maps described in the *Text?* Try one type.

CHAPTER VII

AVERAGES AS TYPES

The Use of Averages in Presenting Wage Statistics[1]

There are two methods of presenting wage statistics: (1) Computation of an average; (2) classification into groups. Each of these methods find frequent illustration in the current literature of wage statistics.

1. *The Average.* — In many instances the only method possible is that of the average, as when the data returned include only the gross amount paid to a given number of workmen. In such a case if a presentation of the wages of the individual be desired, the only available term is an average obtained by dividing the total paid in wages by the number of employees. Such a statistical expression is often valid and instructive, as when the units in the data accumulated are more or less uniform in character and the range of variation is not excessive. At an earlier period when there was greater equality in social and economic conditions, less division of labor, and less variety in industry, the average was relatively a serviceable statistical term; but with the development of modern economic conditions, characterized by the greatest range between skilled and unskilled labor, by many grades of hand and machine labor, and by a multiplication of occupations, the average has become frequently misleading. The advantage of the average is the ease with

[1] Adapted with permission from "Employees and Wages," *Twelfth Census of the United States Taken in the Year 1900*, 1903. Davis R. Dewey, "Report," pp. xxiv–xxviii, Sec. VIII.

which it can be used for formulating a statistical proposition in a single number; it is doubtful, however, whether industrial phenomena so complex as wages can be satisfactorily reduced to a single term. Human labor varies greatly in its form, depending for its effectiveness upon individual skill, intelligence, and energy, as well as upon opportunities for employment. As a result of these variations, rewards differ greatly. Although the economic force of competition exerts a powerful influence toward uniformity of compensation for a given unit of individual exertion as applied in the manufacture of products requiring the same skill and intelligence, yet differences constantly appear; and, as shown by the tables in this report, these differences are found not only within a well-defined occupation in a single section of the country, but even within the same occupation as reported by a single establishment. Some workmen receive high wages, some medium wages, and some low wages; the result is a composite picture, each element of which possesses an individual interest which should not be lost sight of. The student of social conditions is interested to know to how large a part of the social mass certain characteristics, qualities, or phenomena are applicable; and particularly is this true in the study of the condition of labor and its reward. It is far more important to know that one-half of the laboring class receive wages between $1.25 and $1.75 per day, than to know that the average of the total is $1.50. The average disregards the significance of the parts and aims to give expression to the whole in a single term.

2. *Classification into Wage Groups.* — Since variations in wages lose much of their meaning when merged into a single term, the treatment of wage statistics should as far as possible be descriptive, and this is statistically accomplished by the method of classifying wages into groups, as was done, for

example, for certain industries in the Eleventh Census. It must be admitted that this method is not so simple as that of the average; it is much more difficult to compare two lines at all their points than to select from these lines two single points and compare them. For these reasons the method of analysis used in this report for the purpose of comparing wages in different occupations and at two different periods is not as simple as if the average alone had been used. This, however, should not be regarded as a defect; statistical art has its limitations; especially is this so in problems requiring descriptive treatment, such as wages.

An example of the advantage of the classification of wages into groups over the gross average is seen in the following illustration, drawn from one of the pay rolls reported. In this establishment there were 92 employees in 1890 and 299 in 1900. If a general average be desired for all the employees at each of these periods, the results are an average wage of 19 cents per hour in 1890 and 17 cents per hour in 1900, making a reduction of 20 cents per day of 10 hours.[1]

The real difference between 1890 and 1900 is, however, better disclosed in the following table, which classifies the numbers under several rates of wages and also reduces these numbers to percentages of the respective totals for 1890 and 1900.

From this it will be observed that there is a much larger amount of low-priced labor in 1900 than in 1890. Does this mean a reduction in the wages of a given class of employees, as "machinists," for example? The misleading character of a gross average applied to an industry group, as well as the great superiority of a presentation by wage groups such as those in the above table, is disclosed as soon

[1] In computing these averages, the lowest wage in each wage group was taken as the exact wage for each individual in the group.

All employees

Rates Per Hour (Cents)	1900		1890	
	Number	Per Cent	Number	Per Cent
Total	299	100.0	92	100.0
5 to 9	52	17.4	13	14.1
10 to 14	59	19.7	3	3.3
15 to 19	56	18.7	16	17.4
20 to 24	47	15.7	28	30.4
25 to 29	61	20.4	22	23.9
30 to 34	12	4.0	7	7.6
35 to 39	7	2.4	2	2.2
40 and over	5	1.7	1	1.1

as an analysis is made of the several classes of occupations which go to make up the total. Take, for example, the "machinists," of whom 52 were returned in 1890 and 74 in 1900. The distribution of "machinists" according to wage groups is shown in the following table:

Machinists

Rates Per Hour (Cents)	1900		1890	
	Number	Per Cent	Number	Per Cent
Total	74	100.0	52	100.0
15 to 19			5	9.6
20 to 24	10	13.5	19	36.5
25 to 29	47	63.5	20	38.5
30 to 34	9	12.2	6	11.6
35 to 39	6	8.1	1	1.9
40 and over	2	2.7	1	1.9

Obviously the cause of the apparent reduction of wages for all employees is the employment in 1900 of a relatively larger number of low-priced employees than in 1890, probably due to the introduction of improved machinery, which gives a much larger output per machine, but which demands a considerable amount of unskilled labor to handle, erect, assemble, pack, and ship.

Another illustration may be found in an establishment manufacturing fine glazed kid. In 1890 there were 55 employees, all men, and in 1900, 70, of whom 12 were women. The difference in the wages received by males is shown in the following table:

Males in Glazed-kid Factory

Rates Per Week (Dollars)	1900		1890	
	Number	Per Cent	Number	Per Cent
Total	58	100.0	55	100.0
20 and over	2	3.5	1	1.8
15 to 20	2	3.5	3	5.4
12 to 15	6	10.3	14	25.5
10 to 12	6	10.3	15	27.3
9 to 10	3	5.2	4	7.3
8 to 9	17	29.3	5	9.1
7 to 8	14	24.1	3	5.4
6 to 7	4	6.9	4	7.3
5 to 6	1	1.7	4	7.3
4 to 5	3	5.2	2	3.6

It will be observed that there is a marked reduction in the higher-priced labor. This is due to changes which have taken place during the past decade in the manufacture of leather. For example, the reduction in the number of "beamsters" — skilled workmen who remove the superfluous flesh from the hides with a slicking machine — is a result of the introduction of machinery which permits the employment of a greater proportion of unskilled labor. Moreover, the manner of coloring has been changed from table coloring to box coloring; by the former method the color was put on with a brush, whereas now the skins are dipped into a box of coloring liquid. An analysis of the wages of the "beamsters" and the "colormen" does not show any reduction in the wages for the first class.

Rates per Week (Dollars)	Beamsters				Colormen			
	Number		Per cent		Number		Per cent	
	1900	1890	1900	1890	1900	1890	1900	1890
Total . . .	5	10	100.0	100.0	3	9	100.0	100.0
19.00 to 19.49		1		10.0				
15.00 to 15.49						1		11.1
13.00 to 13.49	4	2	80.0	20.0				
12.50 to 12.99	1		20.0					
12.00 to 12.49		7		70.0		3		33.3
11.00 to 11.49					1	4	33.3	44.5
10.00 to 10.49					1	1	33.3	11.1
9.00 to 9.49					1		33.4	

3. *Cumulative Percentage.* — There is one practical defect in classified rates which often impairs their usefulness. This lies in the difficulty of comparing two given sets of returns so as to ascertain what differences may exist or what changes may have taken place; even if the figures in a classified group table be reduced to percentages, the real differences between the two sets of figures are not always easily recognized. For this reason the cumulative percentage has been used in all

Rates per Week (Dollars)	Actual Number at Rate Specified		Percentage in the Group		Cumulative Percentage		Median and Quartile Groups	
	1900	1890	1900	1890	1900	1890	1900	1890
Total . .	759	572	100.0	100.0				
3.50 to 3.99	7	5	0.9	0.9	100.0	100.0		
4.00 to 4.49	10	7	1.3	1.2	99.1	99.1		
4.50 to 4.99	23	15	3.1	2.6	97.8	97.9		
5.00 to 5.49	31	9	4.1	1.6	94.7	95.3		
5.50 to 5.99	12	3	1.6	0.5	90.6	93.7		
6.00 to 6.49	53	40	7.0	7.0	89.0	93.2		
6.50 to 6.99	7	3	0.9	0.5	82.0	86.2		
7.00 to 7.49	22	6	2.9	1.1	81.1	85.7		
7.50 to 7.99	46	37	6.1	6.5	78.2	84.6	Q	
8.00 to 8.49	5	5	0.6	0.9	72.1	78.1		
8.50 to 8.99	1	2	0.1	0.3	71.5	77.2		
9.00 to 9.49	92	42	12.2	7.3	71.4	76.9		Q
9.50 to 9.99	22	6	2.9	1.1	59.2	69.6		
10.00 to 10.49	24	30	3.2	5.2	56.3	68.5		
10.50 to 10.99	60	45	7.9	7.9	53.1	63.3	M	
11.00 to 11.49	25	31	3.3	5.4	45.2	55.4		
11.50 to 11.99	1	5	0.1	0.9	41.9	50.0		M
12.00 to 12.49	100	61	13.2	10.7	41.8	49.1		
12.50 to 12.99	2	3	0.3	0.5	28.6	38.4		
13.00 to 13.49	3	1	0.4	0.2	28.3	37.9		
13.50 to 13.99	75	62	9.9	10.8	27.9	37.7	Q	
14.00 to 14.49	7	4	0.9	0.7	18.0	26.9		
14.50 to 14.99	1	1	0.1	0.2	17.1	26.2		
15.00 to 15.49	62	72	8.2	12.6	17.0	26.0		Q
15.50 to 15.99	13	2	1.7	0.3	8.8	13.4		
16.00 to 16.49	1	1	0.1	0.2	7.1	13.1		
16.50 to 16.99	16	22	2.1	3.8	7.0	12.9		
17.00 to 17.49	2	2	0.3	0.3	4.9	9.1		
17.50 to 17.99	1	1	0.1	0.2	4.6	8.8		
18.00 to 18.49	19	17	2.5	3.0	4.5	8.6		
18.50 to 18.99	1	1	0.1	0.2	2.0	5.6		
19.00 to 19.49	1	1	0.1	0.2	1.9	5.4		
19.50 to 19.99	6	3	0.8	0.5	1.8	5.2		
20.00 to 20.49	4	2	0.5	0.3	1.0	4.7		
20.50 to 20.99		1		0.2	0.5	4.4		
21.00 to 21.49	3	6	0.4	1.1	0.5	4.2		
21.50 to 21.99		2		0.3	0.1	3.1		

Rates per Week (Dollars)	Actual Number at Rate Specified		Percentage in the Group		Cumulative Percentage		Median and Quartile Groups	
	1900	1890	1900	1890	1900	1890	1900	1890
22.00 to 22.49		1		0.2	0.1	2.8		
22.50 to 22.99		4		0.7	0.1	2.6		
23.00 to 23.49		4		0.7	0.1	1.9		
23.50 to 23.99		1		0.2	0.1	1.2		
24.00 to 24.49		3		0.5	0.1	1.0		
24.50 to 24.99					0.1	0.5		
25.00 to 25.49	1	3	0.1	0.5	0.1	0.5		

the detailed tables. The figures in the cumulative percentage column represent the proportion of the total number of persons in the given table receiving a wage as great as, or greater than, the lowest wage of the given wage group. The table above shows the advantages of this method of presentation, and also the method of interpretation.

From this table it is possible to determine how large a proportion of the total number of employees is receiving as much as, or more than, a given wage. For example, the columns headed "cumulative percentage" show that in 1900 the proportion of the total number receiving $8 or more per week was 72.1 per cent, while in 1890 it was 78.1 per cent; at $10 the respective proportions were 56.3 and 68.5 per cent; and at $15 they were 17 and 26 per cent. From the columns of cumulative percentages it is evident that wages were higher in 1890 than in 1900, a fact clearly disclosed neither by the numbers nor by the percentages in the respective groups.

4. *Median and Quartiles.* — The use of the column of cumulative percentages makes it easy to determine the range of wages for any given proportion of the working force; by this means also it is possible to indicate the wage group of

the employee who stands half-way between the lowest-paid and the highest-paid employee in the class under consideration. For example, in the above table, it is seen that when the employees in 1900 are arranged in a sequence according to their rates of pay, beginning with the lowest rate and proceeding upward, the wage of the three hundred and eightieth or middle employee lies between $10.50 and $11.00. The middle term in a series of this character is called the "median." By the use of the median, employees at exceptional rates either low or high, are not given an undue weight or importance as they are when the average is used. Frequently, however, the median will not vary greatly from the average; in the foregoing table, for example, the average in 1900 is $10.55, and in 1890, $11.63.[1] . . .

Another advantage of the cumulative percentage lies in the facility in showing the wages of the employees who stand at selected points along the whole series of employees, as, for example, at one-quarter and three-quarters up the ascending scale. The terms at these particular points are called "quartiles," and within these two limits would clearly fall the wages of at least one-half of the working force. Thus, it will be seen that in 1900 the wages of the employee who stands one quarter of the way up the scale lie in the wage group $7.50 to $7.99; and in 1890, in the group, $9.00 to $9.49. The wages of the employee standing three-quarters of the way up the scale lie in the wage group $13.50 to $13.99 in 1900, and in the group $15.00 to $15.49 in 1890. It is evident, then, that the wages of what may be termed the middle half of the employees were between $7.50 and $13.99 in 1900, and between $9.00 and $15.49 in 1890. Such a statement, however, does not preclude the possibility that more than

[1] In computing these averages, the lowest wages in each wage group was taken as the exact wage for each individual in the group.

one-half of the employees receive wages between the two limits named; it is entirely possible that 60, 70, or a greater, per cent of the working force receive wages within these limits. The method does, however, justify the statement that at least one-half receive the wages stated; there may be more, but there cannot be less.

5. *Limitations in the Use of the Median and Quartiles.* — The limitations in the use of the median and quartiles are of so important a character that they deserve special mention. The use of the median for the comparison of two series of wages is subject to the following drawbacks: The wage scale may be so precise that the tables present data in scattered groups rather than in even distribution throughout the series; then since the median can never fall in any group not represented by actual returns, the change of a few individuals may cause a wide shifting of the position of the median. Or, the groups containing relatively large numbers may be at a distance from the median group, while the group containing the median and the groups near to it may represent only a few persons; in that case also the change of a few individuals about the median rates may appear unduly significant. The shifting of a comparatively small number of persons upward or downward across the median point may thus cause the position of the median group to change in a marked degree. On the other hand the shifting through a considerable distance of comparatively large numbers of persons will not affect the position of the median, provided the median point is not crossed. This is illustrated by the table on page 328.

It will be noted that at both periods there was a combined total of four persons in groups $7.00 to $7.49 and $7.50 to $7.99, while the number of persons both above and below these two groups remained the same (48); and that while the median group was $7.00 to $7.49 in 1890, the shifting of one

Rates Per Week (Dollars)	Actual Number		Cumulative Percentage		Position of Median and Quartiles	
	1900	1890	1900	1890	1900	1890
Total	100	100				
5.00 to 5.49	30	6	100	100	Q	
5.50 to 5.99	10	10	70	94		
6.00 to 6.49	6	30	60	84		Q
6.50 to 6.99	2	2	54	54		
7.00 to 7.49	2	3	52	52		M
7.50 to 7.99	2	1	50	49	M	
8.00 to 8.49	29	9	48	48	Q	
8.50 to 8.99	10	10	19	39		
9.00 to 9.49	9	29	9	29		Q

person upward in the scale made $7.50 to $7.99 the median group in 1900. Yet, although the median advanced a 50-cent group, a heavy fall actually took place in the wages of the majority of the persons shown in the table. The median group would not have changed but for the shifting of one person from group $7.00 to $7.49 to group $7.50 to $7.99. If, instead of the shifting of one of the four persons shown at each period in groups $7.00 to $7.49 and $7.50 to $7.99, the numbers in each of these groups had remained the same at both periods, the median group would not have changed. The median is changed only by a transfer of employees from rates above the median group to rates below it, or vice versa.

The above mentioned defects in the use of the median alone are inherent also in the use of a single quartile, and to some extent in the use of quartiles in pairs. The data at the ends of a scale of wage rates are more likely to be concentrated into isolated groups than those near the center.

6. *Medians with Quartiles.* — The presentation, however, of the median group and the quartile groups together, shows

the change in wages at three equidistant points on the scale, and will as a rule show concisely what the general course of wages has been. Thus, in the foregoing hypothetical example, while the use of the median group alone would have been misleading, a consideration of the median in connection with the quartiles shows that the slight advance in the median group was due to peculiar grouping and scarcity of data at that point, and that there was in fact a considerable fall in wages in the establishment during the decade. Data presenting such irregularity of distribution will more often be found where returns for two or more widely distinct occupations, or different grades of skill in the same occupation, are shown in the same table; with such data, the median and one quartile will often be in the same group. Such a combination might be found in the "total" for an industry, and this possibility affords an additional reason for analyzing wage returns into occupations as specific as possible.

Weighted Averages and Crop Reporting [1]

The numerical method by which the condition of growing crops is measured in Germany is simple in result, but somewhat complex in operation. In the scale adopted 1 represents very good, 2 good, 3 medium or average, 4 poor, 5 very poor. Each correspondent attaches one or the other of these figures to each of the crops reported on, and the averages are worked out in the central office for the whole of Germany. Correspondents are instructed to avoid giving any range as would be implied, for instance, by the use of numbers 1–4, 2–4, 3–5, etc.; where closer estimates are desirable and possible they are permitted to use a decimal point. Thus, if the condition

[1] Adapted with permission from Godfrey, Ernest H., "Methods of Crop Reporting in Different Countries," *Journal of the Royal Statistical Society*, Vol. 73, 1910, pp. 265–266.

of a crop is considered to be midway between 2 and 3, it may be registered as 2.5, and so on for other gradations.

Where there are disturbing factors which prevent the application of a single figure to the whole crop of a district, as, for instance, where a wheat crop on a large area of clay soil may be excellent while that on another area of sandy soil may be the reverse, or where crops differ owing to their cultivation on marshlands, uplands, etc., the correspondent is instructed as to the method he should adopt in order to arrive at a number which fairly expresses the condition of the crop for the whole of his district. He first estimates approximately the area of the crop under each different category, assigns to each the number which properly expresses its condition, and then works out an average figure for the crop in the whole district. The following is a concrete example of the method recommended. Assume that the figure 2 representing "good" expresses the condition of winter rye on marshlands occupying seven-tenths of the whole area of the crop in the district; that 3 or "medium" is the condition of two-tenths of the crop on clay, and that 5 or "very poor" is applied to the remaining tenth on sandy soil, the average condition of winter rye for the whole district will be reckoned as follows:

$$\frac{7}{10}\times 2+\frac{2}{10}\times 3+\frac{1}{10}\times 5=\frac{14+6+5}{10}=\frac{25}{10} \text{ or } 2.5.$$

The yield of a crop in a district of unequal conditions is estimated on the same principle. Thus, assume that the oat crop of a district is divided into seven-tenths on marshlands and three-tenths on sand, that the former proportion yields at the rate of 20 double zentners, and the latter at 10 double zentners per hectare, the average yield of the oat crop for the whole district will be computed as follows:

$$\frac{7}{10}\times 20+\frac{3}{10}\times 10=\frac{140+30}{10}=\frac{170}{10}$$

$$=17 \text{ double zentners per hectare.}$$

The same principle of computation is expected to be applied by the correspondent in cases where the crops have been partly injured by drought, wet, frost, hail, storms, cloudbursts, flood, animal and plant pests, etc., the result being in all cases reported to the Office as the average yield for the cultivated area in the district.

Compensating Errors — The Logic of Large Numbers in Crop Reporting[1]

Crop reports are sometimes called guesses, because they are based upon estimates instead of actual measurements. Of course such estimates are not haphazard guesses; that is, no one would likely estimate the yield of corn at 100 bushels per acre when it is actually only 15 bushels, nor estimate the yield at 15 bushels when it is actually 100 bushels. Nevertheless, nearly every individual estimate has an element of error. Combination of individual estimates into a general average tends to reduce the error in the average. The manner and extent to which this is done may be of interest to the many crop reporters (and others) who frequently feel that their individual estimate may be wide of the truth, but who may not understand fully the effect of combining the estimates of many individuals and thus securing an accurate average.

For the purpose of analysis or study, any error in an individual estimate may be considered as made up of two parts, namely that part which is due to chance and that part which is due to bias.

[1] Adapted with permission from "Monthly Crop Reporter," *United States Department of Agriculture*, March, 1919, p. 31.

A reporter once told us that his father could go through an orchard and estimate its production more closely than any other person in his section, but that he (the reporter) could make a better estimate than his father, after he knew his father's estimate, because he had observed that, although his father made a close estimate, it usually fell under rather than over the final outcome; therefore, by making allowance for this tendency, he could use his father's estimate to make a still closer one.

A bias in an estimate is that part of the error that tends to make it lean more on one side of the actual truth than on the other. The chance error is that part that is equally likely to be above as below the truth.

The chance error in an average of a number of individual estimates tends to decrease as the number of estimates included in the average increases. Suppose any one man's estimate is taken; so far as there is no bias, his estimate is just as likely to be too high as too low or vice versa. Suppose we get an estimate from two men; both may be too high, or both may be too low, or the first may be too high and the second too low; or the first may be too low and the second too high. Observe that there are four possible arrangements. There is one chance in four that both will be too high and one chance in four that both estimates will be too low, but two chances in four, that is, an even chance, that one estimate will be too high and the other too low, thus offsetting each other. If estimates from four men are taken, there will be 16 possible arrangements, and there will be only 1 chance in 16 that all will be too low, but 6 chances in the 16 that there will be 2 overestimates offsetting 2 underestimates. And thus, as the number of estimates taken increases, the chance errors tend to neutralize or offset each other. If only 50 random estimates are obtained and averaged, the probability that all the chance errors will be on the same side (that is, overestimates or underestimates) will be only 1 chance out of 562,949,953,421,312.

If the probable chance error of an individual's estimate is 10 per cent, the probable error of the average of 25 reporters will be only 2 per cent and the probable error of the average of 50,000 reporters will be less than one-twentieth of 1 per cent. An individual may miss the mark as much as 30 per cent, but, in so far as it is equally likely to be too high as too low, the combination of 2500 such estimates (the usual number of returns from county reporters) would give an average which, by the law of averages, would likely be within six-tenths of 1 per cent of accuracy.

It is because of this mathematical law of averages by which large numbers of chance errors in combination tend to offset each other, that the Bureau of Crop Estimates, at small cost as compared with cost of an actual enumeration, can estimate so closely the condition and production of crops.

The bias factor in errors of estimates is more complex than the chance error or guess; it is not eliminated or reduced by increasing the number of reports; it does, however, become more and more nearly constant; and when a biased estimate is compared with a similar biased estimate, the bias is neutralized and thus does not affect the result. For example, suppose the yield per acre of wheat one year is actually 10 bushels, and the reporter, by bias, overestimates the crop 10 per cent; he will report the yield 11 bushels; suppose, again, that the true yield next year is 20 bushels, and the reporter, by bias, continues to overestimate the crop 10 per cent; he will report the yield 22 bushels. It will be observed that the reporter's estimates for the two years, 11 and 22, show the true change; that is, a doubling of the yield, notwithstanding that both estimates were erroneous by 10 per cent. Of course bias is not the same all the time. But the combination of large numbers of reports, obtained from practically the same men and compiled in the same way from month to month and year to year, tends to stabilize the results and make them truly comparable, if not absolutely correct.

REVIEW

1. Compare the discussion with that in Chapter II on *Government Crop Reports*.
2. Restate the case, as developed in the citation immediately above, for the use of the *normal* in crop reporting. What relation, if any, has this discussion to compensating errors as here developed?

The Calculation of the Average Tariff Duty or Rate [1]

It is impossible to compare directly, in any broad way, the rates of duty in different tariff acts. The number of items is large, and while some are of great commercial importance, others are of little importance. The most practical means of comparison is to ascertain the value of imports for all articles or for a group of articles, and also the corresponding amount of duty collected, and, by dividing the amount of duty by the value, compute the average ad valorem rate of duty, or, as it is often called, the average ad valorem duty. This method permits ease of comparison, but, like all averages, has serious defects. Aside from changes in price level, the volume of imports affects the average ad valorem rate just as much as the rate prescribed in the tariff. If, with the same tariff rates in force, one year is marked by specially large imports of goods dutiable at high

[1] Adapted with permission from "Foreign Commerce and the Tariff, 1899–1915," 1916. *Senate Document* No. 366, 64th Congress, 1st Session, pp. 8–9, 13–16.

tariff rates and the succeeding year by specially large imports of goods subject to relatively low tariff rates, the average ad valorem rate of duty will sharply decline. With all its limitations, however, the average ad valorem rate of duty remains the only convenient means of comparing the general level of duties for different years.

In discussing the average rate of duty on imports at different periods, the calculation is made on the basis both of total imports for consumption and of dutiable imports for consumption. For some purposes it is desirable to show the average contribution for all goods entering the country, and this is best disclosed by dividing the amount of duty collected by the total imports. For other purposes it is better to show the level of duties on articles that are dutiable, and for this comparison the duties are divided by the amount of dutiable imports. In making the latter computation account is taken only of the ordinary duties, while the so-called additional duties, varying in amount from $1,198,621 in the fiscal year 1899 to $191,769 in the fiscal year 1915, are excluded. These additional duties represent, in part, the penalty imposed on articles undervalued, which is reported only in the aggregate and not in respect to individual articles, and in part the refund of drawback and the duty, equivalent to internal-revenue tax, collected on articles, grown, produced, or manufactured in the United States when reimported after having been exported. Since these articles are free of ordinary customs duty, they must be excluded from consideration in reckoning the average ad valorem rate on dutiable goods. The additional duties, are, however, included in computing the average ad valorem rate of duty on total imports.

Average Ad Valorem Duty under Recent Tariffs

The average rate of duty on imports under the Underwood-Simmons tariff shows a less marked decrease from the previous rates than is frequently inferred.

The fairest comparison is undoubtedly from the date when the law became effective to the end of the fiscal year, 1914, just one month before the outbreak of the European war. Since the provision admitting wool free of duty did not become effective until December 1, 1913, and the new rates of duty on manufactures of wool did not become effective until January 1, 1914, imports of these articles are included only for the six months, January 1 to June 30. Similarly, imports of sugar and molasses are included only from April 1 to June 30, the first full quarter in which the reduced rates (effective Mar. 1, 1914) were in force. Wool and manufactures of wool are similarly included only for the last two quarters, and sugar and molasses only for the final quarter of the statement covering the nine months ending June 30, 1913.

By this means returns may be compared with no overlapping of tariffs. Throughout the later period the Underwood-Simmons rates were in force, and throughout the earlier period, October, 1912, to June, 1913, the Payne-Aldrich rates were in force.

The average rate of duty from October 1, 1912, to June 30, 1913, was 15.5 per cent ad valorem, calculated on total imports, and 37.8 per cent ad valorem, calculated exclusively on dutiable imports. Similarly, the average rate of duty from October 4, 1913, to June 30, 1914, was 12.3 per cent ad valorem, calculated on total imports, and 34 per cent ad valorem, calculated on dutiable imports. In both periods wool and manufactures of wool are included only for the last six months and sugar and molasses only for the last three months.

The comparison of the results under the two periods shows that for approximately nine months under the present tariff the average rate of duty was 3.2 per cent ad valorem less than under the former tariff, when the average is calculated on total imports, and 3.8 per cent ad valorem less than under the former tariff, when the average is calculated on dutiable imports.

An increase or decrease in the level of duties may be compared on two bases: On the value of imports or on the former average duty. Suppose that all articles imported were subject to a uniform rate of duty of 10 per cent ad valorem and a new law was passed substituting a uniform rate of duty of 8 per cent ad valorem. In such a case the reduction might be described either as 2 per cent ad valorem (that is, 2 per cent of the value of the imports) or 20 per cent of the former duty (2 divided by 10).

The reduction in duty under the present tariff being from 15.5 to 12.3 per cent ad valorem calculated on total imports, means that for the same amount of imports the customs receipts were reduced about 20 per cent of the former duty (3.2 divided by 15.5). Similarly, the calculation on dutiable imports alone shows a reduction from 37.8 to 34 per cent ad valorem. These latter figures indicate that the reduction, considering only the goods that remained subject to duty, represented approximately 10 per cent of the former ad valorem duty. The net effects of the law, so far as shown by the nine-month returns, are therefore a considerable increase in the free list, namely, from 59.2 to 63.8 per cent of the total imports, and a reduction averaging 10 per cent in the rates of duty on the articles that remained subject to duty.

Unfortunately the average rate of duty can not be shown for the six groups of articles classified according to use and degree of manufacture. For the principal items, however, comprising 97 per cent of the total imports, a separation has been made between manufactured articles and unmanufactured articles. The average rate of duty on articles classified as manufactured was 23.8 per cent ad valorem in the nine months ending June 30, 1913, and 19.8 per cent during the nine months ending June 30, 1914. Calculated on the basis of dutiable imports only, the corresponding percentages are 37.1 and 34.4 per cent ad valorem. In the case of unmanufactured goods, the average ad valorem rate of duty on total imports was 6.2 per cent during the nine months ending June 30, 1913, and 4.4 per cent during the nine months ending June 30, 1914. The corresponding figures based on dutiable imports only were, respectively, 41.6 and 32.9 per cent ad valorem. It therefore appears that the decrease in the average rate of duty was much more marked in the case of unmanufactured than manufactured articles. Taking account of total imports, the reduction (4 per cent ad valorem) in the rate of duty on manufactured goods was 16.8 per cent of the former rate of duty on manufactured goods and the reduction (1.8 per cent ad valorem) in the rate of duty on unmanufactured goods was 29 per cent of the former rate of duty on unmanufactured goods. Taking account only of dutiable imports, the reduction in the rate of duty amounted to 7.1 per cent of the former rates in the case of manufactured goods and 20.9 per cent of the former rates in the case of unmanufactured goods.

It is of interest in this connection to compare the effect of the present tariff with that of previous tariffs. The influence of the enactment of the Payne-Aldrich Tariff Act is seen best in a comparison of the returns for the fiscal years 1909 and 1911, which represent, respectively, the last full year of operation of the Dingley Tariff Act and the first full year of operation of the Payne-Aldrich Act. On the basis of total imports the average rates of duty were 23 per cent ad valorem in the fiscal year 1909, and 20.3 per cent ad valorem in the fiscal year 1911, the 2.7 per cent ad valorem decrease representing a reduction of 11.7 per cent of the former rates of duty. On the basis of dutiable imports only, the average rates of duty were 43.1 per cent ad valorem in 1909, and 41.2 per cent ad valorem in 1911, showing a reduction of nearly 2 per cent ad valorem, or about 4.5 per cent of the former rates.

Naturally, the percentage of free goods is larger for unmanufactured than for manufactured goods, under both the Payne-Aldrich tariff and the Underwood-Simmons tariff. For the nine months ending June 30, 1914, under the Underwood-Simmons Act, the average ad valorem duty on dutiable imports was higher for manufactured than for unmanufactured goods, while for the nine months ending June 30, 1913, under the Payne-Aldrich tariff, the ad valorem duty on unmanufactured goods was slightly higher than the corresponding rate for manufactured goods.

This higher average rate on unmanufactured goods was due to the fact that a few unmanufactured articles, of which considerable quantities were imported, had a very high average duty. Thus, during the nine months' period ending June 30, 1913, the average ad valorem rate of duty on tobacco, which constituted nearly one-fourth of the total imports of unmanufactured articles, was 83.77 per cent; wool, which constituted over one-eighth of such imports, had an average ad valorem rate of 44.26 per cent; lead ore had an ad valorem rate of 99.96 per cent; zinc ore, 45.04 per cent; and hay, 55.06 per cent. These articles, except wool, which is now on the free list, had the following average ad valorem rates of duty in the nine months ending June 30, 1914: Tobacco, 82.32; lead ore, 23.88; zinc ore, 10.; hay, 20.44.

A gradual reduction in the average ad valorem rate of duty during a period without tariff change is at first sight a surprising phenomenon. From the fiscal year 1899 to the fiscal year 1909, the period during which the Dingley tariff was in force, the average ad valorem rate of duty on total imports decreased from 29.5 to 23 per cent, and the average ad valorem rate on dutiable imports decreased from 52.1 to 43.2 per cent. There was, thus, under the same tariff, a decrease in the rate of duty of 6.5 per cent ad valorem, based on total imports, and of 8.9 per cent ad valorem, based on dutiable imports, or, in other words, an average reduction, respectively, of 22 and 17.1 per cent of the rates in force in 1899. From 1911 to 1913, the first and the last full years during which the Payne-Aldrich tariff was in effect, there was a similar, though less pronounced, reduction in the average ad valorem rate of duty.

This tendency is due to the gradual rise in prices. Specific rates of duty, which were largely used in both the Dingley and Payne-Aldrich tariff acts, remain unchanged as prices rise or fall, with the result that the equivalent ad valorem rate continually falls as prices rise. The effect of this tendency is obviously to exaggerate the apparent reduction in duty when duties are lowered, and to minimize the apparent effect when they are raised.

The close correspondence between the estimated receipts and the actual receipts under the Underwood-Simmons tariff is striking. It was estimated that the measure, as it passed the House of Representatives, would produce during its first full year of operation $258,000,000; as it passed the Senate, $248,000,000; and as it was finally enacted, $249,000,000. Since the new rates on sugar and molasses became effective March 1, 1914, the law was in full operation only five months before the outbreak of the European war. This covered only one full quarter, that extending from April 1 to June 30, 1914. During that quarter the duties amounted to $63,600, 000, and at this rate the returns for a full year would have been $254,000,000. The receipts, therefore, exceeded by some $5,000,000 the expected proceeds.

Owing to the retention of the old duties on wool, manufactures of wool, sugar, and molasses, it is difficult to make direct comparison for the two quarters ending, respectively, December 31, 1913, and March 31, 1914. The excess receipts on this account during the first quarter under the new tariff may be estimated at $3,600,000 and during the second quarter at $1,100,000. Deducting these amounts from the actual receipts during the two quarters, the revenue, had the act of 1913 been fully in force, would have been approximately $61,700,000 during the quarter ending December 31, 1913, and $65,300,000 during the quarter ending March 31, 1914 — in the one case just about $2,000,000 less and in the other case just about $2,000,000 more than the amount of duty during the quarter ending June 30, 1914, when the new act was in full force.

Averages as Measures of Street Car Utilization.[1]

Utilization of Cars. — Degree of utilization of cars is a relative conception which may be analyzed into several different relations, with corresponding ratios. A certain average number of separate cars is used by a given company more or less in each single day throughout the year. The number of cars in the possession of the company will naturally be in excess of the number used on any particular day; or if rarely

[1] Adapted with permission from *Annual Report of the Public Service Commission of the First District of the State of New York*, 1912, Vol. II, pp. 91–97.

it happens that every car both can be and is put out at some time during the day, the average number for the 365 days will nevertheless necessarily be considerably smaller than the number of cars in the possession of the company, provided the company has a volume of business worth notice. The ratio of the number of cars possessed to the average number used is thus a measure of the reserve supply of cars kept to provide for accidents, repairs, and emergencies of all sorts. But this ratio is subject to qualification with reference to cars designed for use at only one season of the year. Open cars — or strictly speaking, open car bodies, since the same trucks and motors are usually put under closed car bodies in winter — may take the place of most, if not all, the closed cars during several summer months, thus giving an opportunity for thorough overhauling and making it possible to get along with a smaller reserve in winter. If the peak of the demand upon the company comes in summer, however, the open cars may meet that need in a way to make unnecessary a supply of closed cars sufficient to meet the maximum demand of the year. With this qualification, it is the ratio of cars usable all the year, that is, closed cars and convertible cars, to the average number used that gives us a measure of the necessary reserve supply. . . . This is of course a much mixed average for all sorts of transit and cars. It is obviously much affected in the case of two items by the employment of open cars to meet a summer maximum demand. . . .

Another phase of the utilization of cars is reflected by the car miles and car hours operated per car during the year. This comparison can be made either with the cars in the possession of the company or with the average number used. . . . It appears that, in terms of car miles per average number of cars used, the rapid transit lines in general make a considerably greater average use of their cars than do the surface lines, but there are some companies having traffic of an interurban, or almost interurban, character with higher ratios than the rapid transit lines. The ratio of the Hudson & Manhattan is not to be accepted at its face value owing to the unsatisfactory way in which this company determines the "average" number of cars used. In terms of car hours, on the other hand, the degree of use made of rapid transit cars is rather less than the average for the city as a whole. The outlying surface lines, it appears, make a very full use of such rolling stock as they use at all, that is, the proportion of cars used to cars owned is low, but that of car miles to cars used is high. All these ratios, however, are somewhat subject to qualification, owing to the lack of sharp definition of the average number of cars used per day and of careful conformity to it on the part of the companies. . . .

Seat mileage operated is not, or should not be, the crude product of the average seating capacity times the number of car miles operated, and therefore the dividend obtained by reversing this process, that is, the ratio of car miles operated to seat miles operated, should not be expected to coincide with the average seating capacity of cars owned. The contribution of open cars to car mileage will obviously not be in proportion to their number, since they are operated only during the summer months, while it will be considerably more than in proportion to their car mileage, owing to their large size, in terms of seating capacity. . . . .

But the two ratios are near enough together to indicate the substantial accuracy of the seat-mile return. But an incorrect attribution of seating capacity to cars in the first instance would affect both similarly. The fact that the average seating capacity of cars in the possession of the companies is a trifle smaller than of cars operated may be explained by a preferential use of the newer and on the average larger cars. On the other hand, the open cars, with their larger seating capacity, should have more influence upon the average for cars owned or leased, owing to their use in summer only. This factor seems to have been counterbalanced by the one just referred to. . . .

The greater the seating capacity the fuller is the utilization by a company of its individual cars. An appreciation of this fact accounts for the tendency of street railways to use larger cars. Traffic conditions, however, limit the possibility of taking advantage of this economy. For some types of service, moreover, facility in loading and unloading is more important than additional seats.

REVIEW

1. With what types of units does this article deal? Consult the *Text*, Chapter III, and Professor Bowley's notion of "relativity" in *The Nature and Condition of Statistical Measurement*, in Chapter III, *supra*.

2. What is the "measure of the necessary reserve supply" of cars? Why is this a "much mixed average"?

3. What conditions would affect a comparison of car hours, and car miles on a given line, and on different lines?

4. Define the unit seat mile. How may the number of seat miles be calculated for a given line? What effect has the greater use of new cars, and of the larger cars on this average?

Car-Seat Mile Averages and Ratios [1]

Car-seat Mile Ratios. — Ratios of seat miles to passengers are better comparable as between companies and between different years than car-mile ratios, since allowance is made for difference in the seating capacity of cars. The table be-

[1] Adapted with permission from *Annual Report of the Public Service Commission of the First District of the State of New York*, 1913, Vol. II, pp. 76–78.

Seat Miles in Relation to Passengers and to Car Miles, 1912 and 1913

Roads	Seat Miles to Passengers			Seat Miles to Car Miles		
	Ratios		Points difference between ratios	Ratios		Points difference between ratios
	1912	1913		1912	1913	
Hudson & Manhattan	5.65	5.79	+0.14	44.00	44.00	
Interborough	10.66	10.28	−0.38	49.95	49.97	+0.02
Rapid Transit subway	10.85	10.21	−0.64	52.00	52.00	
Manhattan elevated	10.47	10.35	−0.12	48.00	48.00	
Brookyn Rapid Transit	8.53	8.29	−0.24	47.23	47.38	+0.15
Elevated division	9.98	9.79	−0.19	52.12	52.11	−0.01
Surface division	7.63	7.39	−0.24	43.89	44.19	+0.30
Bridge Locals	2.41	2.34	−0.07	34.99	33.67	−1.32
Brooklyn bridge	2.36	2.24	−0.12	35.95	36.00	+0.05
Williamsburg bridge	2.30	2.01	−0.29	37.63	37.63	
Queensboro bridge	2.91	3.05	+0.14	28.00	28.00	
Manhattan bridge		3.47			28.00	
Manhattan surface	5.75	5.39	−0.36	40.11	40.98	+0.87
Electric contact	5.79	5.43	−0.36	41.55	42.54	+0.99
Storage battery	5.25	4.98	−0.27	21.87	22.86	+0.99
Horse	4.90	4.42	−0.48	23.22	22.60	−0.62
Bronx Surface	9.74	9.69	−0.05	45.24	45.87	+0.63
Trolley	9.80	9.74	−0.06	45.29	45.91	+0.62
Monorail electric	3.12	3.27	+0.15	43.96	43.84	−0.12
Horse	1.61	2.49	+0.88	17.18	21.52	+4.34
Brooklyn, Excl. B.R.T.	9.84	9.25	−0.59	48.60	47.56	−1.04
Queens, Excl. B.R.T.	9.66	8.92	−0.74	43.52	43.42	−0.10
Richmond	8.94	9.03	+0.09	37.79	39.25	+1.46
Underground	10.01	9.53	−0.48	51.16	51.14	−0.02
Elevated	10.29	10.15	−0.14	49.37	49.37	
Total rapid transit	10.17	9.87	−0.30	50.11	50.11	
Conduit electric	5.79	5.43	−0.36	41.57	42.55	+0.98
Trolley	8.12	7.84	−0.28	44.05	44.31	+0.26
Storage battery	5.25	4.98	−0.27	21.87	22.86	+0.99
Monorail electric	3.12	2.90	−0.22	43.96	38.82	−5.14
Total electric surface	7.05	6.74	−0.31	42.94	43.46	+0.52
Horse	4.86	4.39	−0.47	23.18	22.59	−0.59
Grand total	8.58	8.26	−0.32	46.66	46.91	+0.25

Ratios for grand totals of prior years

	Earlier year	Later year	Difference	Earlier year	Later year	Difference
1911 and 1912	8.60	8.58	−0.02	46.60	46.66	+0.06
1910 and 1911	8.47	8.60	+0.13	46.28	46.60	+0.32

low gives such ratios, as well as ratios of seat miles to car miles, for the main groups of companies.

The most striking feature of the table is the decrease in the ratio of seat miles to passengers which took place in 1913 in the case of nearly every group shown in the table. This is due of course to the smaller increase in accommodations than in passengers, which has already been noted, and is of advantage to the companies and likely to be to the disadvantage of the traveling public. The Queens roads profited most in this respect, and the Interborough subway was next. In 1912 the latter gave the greatest service in exchange for a nickel as measured by seat miles, but in 1913 the amount of such service was surpassed by that offered by the Manhattan elevated. To one learning the fact for the first time, it appears surprising that the most congested of all lines, the Manhattan and Brooklyn elevated and the Interborough subway, should give the greatest number of seat miles per passenger. The high ratios are of course due to the unusually long average ride taken by passengers and to the immense number of empty seats during the last mile or two of the trip to the outskirts of the thickly settled portion of the city. The small ratios of the bridge locals, the monorail and the Bronx horse cars, are of course due to the shortness of the route. . . .

The ratios of seat miles to car miles are equivalent to the average seating capacities of cars actually in use, as distinguished from the average seating capacity of all cars owned or leased. . . . For several years the average capacity has continuously been increasing for the city as a whole, due both to the increasing proportion of rapid-transit traffic, which employs cars of comparatively large seating capacity, and to the installation of new and larger cars on the surface lines. The most marked increases in seating capacity since 1910 — the first year for which figures for seat miles are available — are shown for the Manhattan and Richmond surface roads, 6.6 per cent, which is almost equaled by the increase for Bronx surface roads, 6.5 per cent. The average capacity has not changed at all during the 3-year period for the Hudson & Manhattan, Interborough subway, Manhattan elevated, and Queensboro bridge locals. It has slightly decreased for the Brooklyn Rapid Transit elevated and surface, the other Brooklyn roads, the Queens roads, and the Williamsburg and Brooklyn bridge locals. In the case of the Brooklyn and Queens surface roads, the decrease is probably due to a decrease in the proportionate use of open cars, which have a considerably larger seating capacity than closed cars of the same size. For the Brooklyn elevated lines on elevated structures, the average seating capacity slightly increased, the decrease for the elevated division as a whole being due to the change on the South Brooklyn and Sea Beach lines on the surface over which "elevated" trains run.

REVIEW

1. Put into the form of a general statement the conditions that should be observed in comparing the seat miles of two different lines.

2. How is this discussion related to the contention of the *Text* that "like can be compared only with like"?

3. What conditions have operated to change the ratios of seat miles to car miles?

REVIEW PROBLEMS

Averages

1. Using the data in the table below

(1) Compute the arithmetic average expenditure for breakfast, dinner, and supper.

(2) Compute the median expenditure — to the nearest group and also to the nearest cent — for breakfast, dinner, and supper.

Table Showing the Expenditures for Food by Men and Women and by Meals

Expenditure Groups (cents)	Meals and Purchasers of Food											
	Total			Breakfast			Dinner			Supper		
	Total	Men	Women	Total	Men	Women	Total	Men	Women	Total	Men	Women
Total	6843	2897	3946	836	359	477	3233	1391	1842	2774	1147	1627
3 to 7	15	7	8	5	1	4	6	2	4	4	4	—
8 to 12	188	64	124	84	25	59	57	12	45	47	27	20
13 to 17	516	150	366	252	91	161	183	39	144	81	20	61
18 to 22	763	230	533	220	87	133	356	98	258	187	45	142
23 to 27	982	343	639	134	65	69	552	186	366	296	92	204
28 to 32	849	345	504	70	42	28	497	211	286	282	92	190
33 to 37	672	315	357	34	25	9	350	179	171	288	111	177
38 to 42	758	334	424	19	11	8	336	174	162	403	149	254
43 to 47	702	351	351	14	12	2	297	164	133	391	175	216
48 to 52	563	307	256	2	—	2	224	120	104	337	187	150
53 to 57	407	223	184	1	—	1	159	89	70	247	134	113
58 to 62	179	106	73	—	—	—	77	47	30	102	59	43
63 to 67	100	49	51	—	—	—	60	29	31	40	20	20
68 to 72	75	37	38	—	—	—	38	23	15	37	14	23
73 to 77	36	16	20	—	—	—	18	7	11	18	9	9
78 to 82	17	11	6	—	—	—	9	5	4	8	6	2
83 and over	21	9	12	1	—	1	14	6	8	6	3	3

(3) Locate the modal expenditure — to the nearest group and also to the nearest cent — for breakfast, dinner, and supper.

2. Compare the averages (arithmetic means) medians, and modes.

(1) Arrange these in the form of a table. Give the same a proper title and express comparatively in the table the relations which they bear to each other.

(2) How nearly is the contention in the *Text* realized, that these averages for series, not too asymmetrical, stand in a definite relationship?

(3) How differently, if at all, would you interpret these averages if the series were continuous rather than discrete?

3. By the use of the averages computed in this problem, verify the properties of averages as described on pp. 279–289 of the *Text*. How satisfactory would it be solely to speak of these expenditures in terms of averages?

4. Using the data above, but reduced to percentages,

(1) Compare, by using simple *percentages*, frequency graphs drawn on a single figure and to a common scale, the expenditures for breakfast, dinner, and supper.

(2) Locate the mode graphically and compare your figure with that determined arithmetically in Problem 1 — (3).

(3) Indicate on the graphs the positions of the medians and arithmetic means, as determined in Problem 1 — (3). What order do they have? Are they equally distant apart? Absolutely? Test by reference to Problem 2. Express the relations graphically by the use of bar diagrams.

5. From your answers to Problems 1–4, and from such other computations as seem to you to be necessary, answer the following questions:

(1) Are the men more or less consistent than the women in their expenditure for different meals?

(2) In their expenditure for all meals?

(3) How do you measure consistency?

(4) Do your graphs in Problem 4 help you to answer these questions? In what way, if at all?

6. By applying an entirely different set of weights from those used by the Bureau of Labor, see p. 176, calculate, for the same accidents, both a severity rate and a frequency rate. What effect seems to be assignable to the weights? Do you agree with the generalization that "the character of the weighting scale used becomes comparatively unimportant"? Does your one illustration serve as an adequate basis for giving an answer? Try other weights.

CHAPTER VIII

PRINCIPLES OF INDEX NUMBER MAKING AND USING

METHOD OF COMPUTING INDEX NUMBERS — BUREAU OF CROP ESTIMATES [1]

THE trend of prices to farmers for important crops is indicated in the following figures; the base, 100, is the average price December 1 in the 43 years 1866–1908, of wheat, corn, oats, barley, rye, buckwheat, potatoes, hay, flax, and cotton:

	1919	1918	1917	1916	1915	1914	1913	1912	1911	1910
Jan. 1	272.4	264.1	183.6	129.0	126.7	132.5	110.9	133.9	118.6	134.1
Feb. 1	259.9	271.6	195.6	139.9	140.5	132.1	112.6	140.2	119.8	138.5
Mar. 1	257.1	288.8	206.5	138.6	144.0	133.8	113.3	144.7	117.9	139.9
Apr. 1	271.2	288.6	225.2	140.2	144.5	134.2	113.6	153.4	118.0	138.8
May 1	293.7	281.8	280.6	143.3	150.0	135.9	116.2	166.3	122.2	133.5
June 1	307.2	271.9	291.3	145.8	147.3	138.8	121.2	168.3	127.7	133.5
July 1	310.2	272.9	289.9	144.8	139.1	137.7	122.9	160.1	136.3	133.1
Aug. 1	—	280.6	307.8	147.7	138.9	137.6	125.4	148.0	148.2	137.1
Sept. 1	—	293.3	279.6	161.5	132.5	141.3	136.3	137.6	141.6	137.0
Oct. 1	—	289.3	277.0	163.6	128.2	136.4	139.1	128.6	138.0	129.8
Nov. 1	—	266.5	261.3	178.8	124.4	127.4	133.9	118.3	135.6	122.2
Dec. 1	—	265.5	252.3	187.9	120.4	122.8	132.7	110.3	133.1	118.4

The index numbers of prices as published by the Bureau of Crop Estimates of the United States Department of Agriculture, are the result of —

[1] Taken with permission from "Monthly Crop Report," *United States Department of Agriculture*, July, 1919, and August, 1918, pp., respectively, 67 and 96.

(*A*) A comparison of the current price of each of 10 crops with its average December 1 price for the 43-year period 1866–1908, and

(*B*) A combination into one figure of the 10 index numbers thus obtained for the 10 crops, by weighting them with figures approximately proportionate to the importance of the several crops in the aggregate value of the 10 crops for a series of years.

These processes may be shown as follows:

A

	(1) 43-Year Average Dec. 1 price	(2) Current Price (Apr. 1, 1917)	(3) Index Number (Column 2 Divided by Column 1 and Multiplied by 100)[1]
Wheat	$0.8450	$1.800	213.0
Corn	.4148	1.134	273.4
Oats	.3274	.615	187.8
Barley	.5747	1.023	178.0
Rye	.6271	1.356	216.2
Buckwheat	.6228	1.283	206.0
Potatoes	.5276	2.347	444.8
Hay	9.3820	13.050	139.1
Flax	1.0000	2.661	266.1
Cotton	.1000	.180	180.0

The following tabulations show the different steps in obtaining the final index number, in their logical order. In office work, however, a much simpler process is used, as the result of uniting into a combination weight, or constant (the same for all months), the various known factors, leaving only the single unknown factor (current price) to

[1] That is, per cent that current price is of 43-year average Dec. 1 price.

B

	Index Number	Weight (Approximate Proportion of Aggregate Value of 10 Crops represented by each Crop)[1]	Extension
Wheat	213.0	176	37,488.0
Corn	273.4	325	88,855.0
Oats	187.8	93	17,465.4
Barley	178.0	28	4,984.0
Rye	216.2	6	1,297.2
Buckwheat	206.0	3	618.0
Potatoes	444.8	55	24,464.0
Hay	139.1	172	23,925.2
Flax	266.1	7	1,862.7
Cotton	180.0	135	24,300.0
10 crops combined . .	[2]225.2	1,000	225,259.5

be applied when determined at the time of the report. This method of simplification by factoring may be shown as follows:

Representing current price by P and the crops by small initial letters and analyzing operations called for in tabulations above, we have —

$$176 \times 213.0 = 176 \times \frac{Pw}{.8450} = \frac{176}{.8450} \times Pw = 208 \times Pw$$

$$325 \times 273.4 = 325 \times \frac{Pc}{.4148} = \frac{325}{.4148} \times Pc = 783 \times Pc$$

$$93 \times 187.8 = 93 \times \frac{Po}{.3274} = \frac{93}{.3274} \times Po = 284 \times Po, \text{ etc.}$$

[1] Obtained by multiplying 1909 production for each crop by 43-year average price and dividing the resultant product by the aggregate of such values (regarded as the base, or 1000) for the 10 crops.

[2] Extension divided by weight.

Constants having been thus obtained once for all, the operation to be performed at the time of the report is thereby condensed into the simple operation of multiplying current prices of the individual crops by their respective constants and pointing off the sum of the extensions.

The sum of the extensions is practically the same as the sum of the extensions in tabulation B — they are identical if both operations are carried out to the same degree, no additional factors having been included in tabulation C; therefore the index number is, as in tabulation B, the sum of the extensions divided by 1000, or 225.2.

For April 1 the results were as follows:

C

	Combination Weight, or Constant	Price April 1, 1917	Extension
		Cents	
Wheat	208	180.0	37,440.0
Corn	783	113.4	88,792.2
Oats	284	61.5	17,466.0
Barley	49	102.3	5,012.7
Rye	10	135.6	1,356.0
Buckwheat	5	128.3	641.5
Potatoes	104	234.7	24,408.8
Hay	18.3	1,305.0	23,881.5
Flax	7	266.1	1,862.7
Cotton	1,350	18.0	24,300.0
10 crops combined			[1] 225,160.4

The ten crops considered in the index number comprise nearly 90 per cent of the area in all field crops, the average value per acre of which closely approximates the value per acre of the aggregate of all crops. Therefore, the index

[1] Index number, 225.2.

numbers based upon these crops may be regarded as practically the same as if all the minor crops were included. The December 1 price for 43 years, 1866–1908, was used because it was the longest period of prices available when the index numbers began, in 1908.

The Why and How of Stock Index Numbers [1]

In recent years index numbers of stock prices have gained general acceptance: they are regularly "carried" by the financial press; they are watched by bankers, investors, and speculators; they are put before railway commissions and courts as evidence; they are used in many ways by publicists and economists. This acceptance, however, is not the result of critical approval. Perhaps the good repute which index numbers of commodity prices at wholesale have fairly won after long discussions has disposed most "consumers of statistics" to trust index numbers as such. But apart from any special justification, there certainly prevails an amiable willingness to take upon faith plausible figures that fill a pressing want. And the stock index numbers have been published in the form that makes new figures most alluring — the paucity of explanations and warnings has encouraged readers to use or misuse the results without undergoing the mental toil of criticism or the moral strain of doubt. As for the cautious minority, they have been foiled by this same simplicity of presentations; they have been given few materials wherewith to determine the representative value of the original quotations, to judge the appropriateness of the methods used, or to compare the results of rival series. . . .

[1] Adapted with permission from Mitchell, Wesley C., "A Critique of Index Numbers of the Prices of Stocks," in *Journal of Political Economy*, July, 1916, Vol. XXIV, pp. 625–631.

The Fundamental Difference between Stock and Commodity Index Numbers. — In several respects stock prices are more satisfactory data for statistical analysis than commodity prices. Stock dealings are more highly centralized and more thoroughly organized than dealings in most commodities. The prices are reported with unexcelled fullness and accuracy. While the number of stocks for which frequent and regular quotations can be collected for considerable periods is less than the corresponding number of commodities at wholesale, it doubtless forms a larger proportion of the whole list dealt in. Once more, stocks are quoted in terms of a nominally uniform unit — the share with a par value of $100, or some multiple that can readily be changed into the standard unit. Hence the actual prices can be compared, summed, and averaged with a facility lacking when one handles commodity prices. Concerning the authenticity and the representative character of stock quotations, in short, there are fewer doubts than haunt the mind of the field-worker in a commodity-price investigation.

It is when one begins to interpret these quotations that doubts become grave. First there is the familiar question: What does the share with a par value of $100 really mean? Second, there is the assurance that whatever that unit in one corporation means this year, it will probably mean something different next year. Commodities are tangible substances, measured by physical units, and in making index numbers one rejects articles that are not substantially uniform in quality over long periods. Business enterprises, on the contrary, are essentially variable entities, and shares in them are subject to changes that affect the enterprises, and to other changes as well. The Pennsylvania Railroad, for example, is a remarkably stable corporation; yet its physical property, its security holdings, its leases, its indebtedness, its earnings and expenses, its financial affiliations, its relations to regulating commissions, and a hundred other matters that affect the market value of its shares, all vary constantly or intermittently. To cite only the one crude gauge: the Pennsylvania system counted about 7600 miles of railway in 1890 and about 11,800 miles in 1915. And in this changing property a share of common stock in 1890 represented ownership of one part in 2,451,354, whereas in 1915 one share represented ownership of one part in 9,985,314. Stocks, then, are variable fractions of variable wholes, and their prices fluctuate incessantly because of changes in the *thing quoted,* as well as for other reasons.

From such facts it is sometimes inferred that index numbers of stock prices have no valid use except for short-period comparisons; or that an index number covering decades is no better when it excludes than when it admits numerous substitutions of one stock for another. In any case, the argument runs, comparisons of stock prices in years far apart are comparisons of dissimilar goods; they are like comparisons of the prices of potatoes and silk in 1890 with the prices of pig iron and tea in 1915.

Such conclusions, however, are rash. The fact that stocks change as commodities do not, proves merely that stock index numbers must not be interpreted as meaning precisely what commodity indexes mean. It does not prove that stock indexes are meaningless, or that alterations in the list of securities included in them are unobjectionable. Business enterprises, indeed, are more like men than they are like commodities. Commodities are produced and consumed; then produced afresh in the old forms. Business enterprises have a continuous life; they undergo great changes of expansion, contraction, even reorganization, without losing their identity. And this continuity of business enterprises and of shares in them is a fact of great practical importance. The many individuals and corporations that hold stock in the same business enterprises for years at a time are deeply concerned with long-period changes in the prices of their securities. The like holds true with reference to the "investing public" as a unit, and to its security holdings as an aggregate. Even the wider public in its efforts to regulate corporation charges and corporation finance through governmental commissions needs to know the course followed by security prices in particular and in general. The fluctuations of New Haven stock between the early nineties and the present are not rated a matter of indifference; neither are the very different fluctuations of Pennsylvania stock, nor the still different fluctuations of Lackawanna. Nor is it unimportant to find out which type of fluctuations has been characteristic of American stocks at large.

Stock indexes, then, differ from commodity index numbers in that they show, not variations in the prices of unvarying goods, but variations in the prices of goods that maintain their identity despite continual changes in quality. This difference enhances the difficulty both of making and of using them; but it does not destroy their logical legitimacy or their practical importance. . . .

The Uses of Stock Index Numbers. — An index number is a statistical device made to serve certain ends. Hence the logical first step toward an evaluation of any such series is to define precisely the end which the finished results are to serve. That done, one has a criterion by which to judge the merits and defects of the series already in use, and by which to guide his own efforts in making new series.

The trouble with this seemingly promising lead is that stock index numbers are put to so many and such varied uses as to give little help in defining what is wanted. An economist may seek to measure changes in the purchasing power of money over stocks, a speculator may wish to forecast the probable future course of the market, a public commission may be interested in the terms on which corporations can raise new capital, a publicist may investigate the claim that government regulation has brought loss upon investors, a financial historian may wish to mark off periods of expansion and contraction, a trustee may inquire whether the fluctuations of his security holdings have compared favorably with the average course of the market, an insurance company may seek light on the probable future of interest rates, a student may wish to compare stock fluctuations with the price fluctuations of commodities of wholesale or retail, of labor, of bonds, of farm lands, of securities in other countries, etc. Now, each one of those people will have use for a stock index. But the more carefully these various uses are analyzed the clearer it becomes that their requirements differ. The character and the number of stocks to be included, the frequency of the quotations needed, the period of time covered, whether actual or relative prices should be used, the desirability of making subgroups and their basis, the kind of average appropriate, the necessity of considering deviations from the mean, whether weights should be introduced and if so what is the proper criterion of "importance" — these and the other points of technique that arise in making an index number would not all be decided precisely alike in any two of the cases suggested, did uses strictly dictate methods — as logically they should.

Ideally, every distinct use should have a distinct index

number especially designed for it. Practically, however, cases are few when the consumer of statistics has the technical skill and can spend the time and money to make a series exactly answering his needs. What happens is that he uses for his special purposes one of the series published by others — more often than not without realizing that the figures in question are in certain respects ill adapted to his needs. Frequently the user does not even hit upon that one among the published series which is least unsuited to his case. And this situation promises to change but slowly. Probably the published series will long continue to be used as "general-purpose" index numbers. And a "general-purpose" index number is too indefinite a conception to guide one surely through the maze of choices that are involved in making a new series or in ranking old ones.

Under these confusing circumstances, what can we attempt with any prospect of success? We cannot discuss the merits of stock index numbers at large with reference to their uses, because these uses and their several requirements are so multifarious. Our best hope seems to lie in reversing the problem. That is, we can analyze stock index numbers, old and new, to find out of what materials and by what methods they are made. Then we can discuss their uses with reference to their construction. Finally, we can determine what fluctuations in the prices of stocks can be measured most accurately and by what means. The index number which stands first in this test will have special claims to acceptance, except for uses which require some radically different series, less accurate though it be.

REVIEW

1. Contrast commodity and stock prices in relation to index number making and using.

2. What is Professor Mitchell's approach to a study of stock indexes? In what way does he contrast general- and special-purpose numbers? In what ways is his discussion paralleled in the *Text*?

Weighting and the Making of Stock Index Numbers [1]

So long as statisticians expected but rough results from their index numbers of commodity prices at wholesale, they treated systematic weighting as a theoretical refinement in method which made little difference in the results. What pleased them was to find that their simple and weighted averages showed the same general trend. But as experience has demonstrated that under favorable circumstances the margin of uncertainty in such work may be reduced to less than, say, 10 per cent of the results, makers of commodity index numbers have begun to regard proper weighting as practically important. Is it important in also making index numbers of stock prices?

Hitherto most stock indexes have been "simple" averages of actual or relative prices. Now simple averages into which no weights enter, or in which all stocks have the same weights, — they are really averages in which the weights have not been systematically planned but left to chance. What degree of influence any stock in a given sample will exercise upon the results in a simple series depends both upon the original quotations and upon the way in which they are worked up. For example, an arithmetic mean of actual prices in effect assigns heavy weights to the stocks that command high prices per share and light weights to stocks that are cheap. But if these actual prices are turned into relatives and the arithmetic means are made from the

[1] Adapted with permission from Mitchell, Wesley C., "A Critique of Index Numbers of the Prices of Stocks" in *Journal of Political Economy*, July, 1916, Vol. XXIV, pp. 684–691.

latter figures, the weighting is likely to be revolutionized. For now the influence of a given stock depends on a radically different factor, not on its price in dollars and cents as compared with the prices of other stocks in the sample, but on the percentage which the price on the date in question bears to the price of the same stock in the period chosen as base as compared with the corresponding percentages for the other stocks. A shift to a new base commonly alters the relative magnitude of these percentages and therefore changes the weights once more. Finally, the substitution of geometric means or medians for arithmetic means gives an entirely new twist to the whole situation. In a geometric mean the influence of a stock depends upon the comparative magnitude of the *ratios* of change which its price undergoes, and it matters not whether actual or relative prices are used or on what base the relative prices are computed, for none of these matters affect the ratios of change, which alone count. In a median it does make a difference whether actual or relative prices are averaged and on what base the relatives are computed; but the influence which any stock exercises upon the result depends solely on whether its actual or relative price happens to be at, above, or below the middle of the whole series after the data have been arranged in numerical order. The magnitude of its deviation from the middle position has no effect.

Since all index numbers are really weighted, the only question is whether these weights should be tacit or avowed, obscure or clear, left to chance or controlled on some intelligible principle. This question is one of great moment, particularly when one is dealing with stocks. For different schemes of systematic weighting produce large differences in results, when the weights themselves differ notably. Different schemes of haphazard weighting tacitly introduced by changing from averages of actual prices to averages of relatives, or by shifting the base on which relatives are computed, cause wide divergences. Finally, in most cases the series with systematic weights and the series with haphazard weights differ from each other at least as much as they differ among themselves. If systematic weighting is desirable in making commodity indexes where it leads to comparatively moderate differences in results, a fortiori it is desirable in making stock indexes where the differences produced in results are much wider.

Few men would hestitate to say that the price of Pennsylvania stock is more important than the price of Duluth, South Shore & Atlantic stock and deserves to have more weight in an index number. It is more important because there is more Pennsylvania stock in the hands of investors, individual and corporate; because the Pennsylvania does much the bigger business; because Pennsylvania stock is a more important article of commerce — more of it changes hands year by year.

These three reasons imply three different criteria of the importance of a given stock, criteria upon which may be based three sets of weights, each of which is appropriate for special ends. If the aim is to show the average changes in the prices of securities held by the public, the amount of stock outstanding yields the logical set of weights. If the aim is to throw light on the changes in the prices of business enterprises as such, then gross earnings, the best available gauge of volume of business transacted, may be used as weights. If the aim is to find average changes in the prices of stocks that are traded in, then the number of shares sold should be used. Other aims might make still other systems of weights desirable. . . .

But to what ought these weights be applied — to actual prices or to relative prices worked out on some chosen base? That is equivalent to the question: What weights ought to be used on the actual prices? For any average of relative prices is itself a weighted average of actual prices in disguise. For example, index numbers made by averaging relative prices on the 1890–1899 base are equivalent to averages of actual prices weighted by the factors required to make the average actual price of each stock in that decade equal 100. . . . Similarly the index numbers of relative prices on the preceding year base are averages of actual prices, each weighted by the multiplier, which makes its price in the year before equal 100.

In weighting relative prices, then, we are weighting already weighted actual prices. Upon the final result, therefore, each stock will have an influence proportioned, not to its figure in the formal scale of weights, but to this figure combined with its actual-price-times-another-weight. Likewise, in weighting actual prices themselves we give each stock an influence upon the result which depends, not simply on the weight, but upon the product of the weight times the price. . . .

The first step in weighting, therefore, should be to decide what proportionate influence we wish each stock to exercise upon the final results. Of course that depends upon the end in view. For example, in measuring the changes in the market value of stocks held by investors, the importance of each of our sample stocks depends both on the amount in the hands of investors and on the actual price. Weights based on amounts outstanding should therefore be applied to actual prices. If for some other purpose we think that the fluctuations of each stock should have an influence proportionate simply to the gross earnings of each corporation, then we should not apply weights based upon earnings directly to actual prices, but should first make the average actual prices of all the stocks the same for the period covered by applying one set of equalizing weights, and then multiply these equated prices by the weights based upon earnings. In this case, however, it would be quicker to begin by making the two sets of weights into one, and then to multiply the actual prices of the stocks by the consolidated weights.

REVIEW

1. Professor Mitchell seems to distinguish between the theoretical and practical aspects of weighting. State his distinction, and compare it with the discussion of the same subject in the *Text.*

2. How differently do weights operate in simple averages of actual prices, and of relative prices? How differently in the case of medians; in the case of geometric means?

3. Defend the writer's contention that "all index numbers are really weighted."

4. What criteria of importance may be used in selecting weights for stock indexes? Under what conditions should each be selected?

5. To what form of the price data ought weights to be applied? Why is this an important question?

Conclusions on the Making of Stock Index Numbers [1]

The choice of methods in making an index number of stocks should be guided by the specific purpose in view. It follows that the index number that is best for any purpose depends upon the specific phase of price fluctuations which that purpose requires to be measured.

[1] Adapted with permission from Mitchell, Wesley C., "A Critique of Index Numbers of the Prices of Stocks" in *Journal of Political Economy*, July, 1916, Vol. XXIV at pp. 691–693.

Strictly interpreted, this obvious but often-neglected rule bars out the question: What is the best index number at large? Perhaps there is no single series that is not "the best" for some imaginable use. But, by way of conclusion, we may point out what fluctuations in the prices of stocks can be measured with the narrowest margin of error, and argue that the index number which best represents these most measurable fluctuations is the best "general-purpose" series; the index number to be recommended for use by the general reader, and by the specialist also, when his particular aim does not definitely demand some differently constructed series, in spite of its inferior accuracy.

Along this line a confident opinion can be given. Geometric means of the ratios of change in quotations within brief periods, such as from one year to the next, have been shown to be the most accurate measures of fluctuations in the prices of stocks. . . .

For measuring fluctuations covering longer periods of time geometric means are again the most representative averages. But the farther apart grow the years between which price comparisons are made the less accurate grow the results obtained from a given body of quotations and the smaller grows the list of stocks for which continuous series of quotations can be had. It is true that the successive percentages of change in price from one year to the next can be multiplied into each other to make a continuous "chain index;" but, while each link has a narrow margin of error, the errors are cumulative, so that a comparison between the two ends of the chain becomes less trustworthy the longer the chain is made. Of course the same difficulty inheres in the relative prices on a fixed base that may be made from the geometric means of actual prices. No refinement of methods can mend the fundamental defect of the data. The ratios of change in stock prices between years far apart are so widely and irregularly scattered that no average made from them can have a high representative value.

The best way to diminish, since we cannot remove, this difficulty is to break the long periods up into parts, to compute fresh index numbers for each part, and to string these index numbers together. The advantages of this shift are (1) that a larger "sample" of stocks with continuous quotations can be had for short periods, and (2) that the fixed-base relatives will show a less irregular distribution. Pushed to extremes, this course would lead to the making of a geometric-mean index number of all stocks quoted both in 1890 and in 1891, and of a second index number of all stocks quoted both in 1891 and in 1892, and so on to date. The main defect of such a series, after the yearly percentages had been linked together in a chain index, would be that no one could be sure what part of the fluctuations shown was due to change in prices and what part to changes in the stocks quoted. Hence price comparisons between 1890 and 1915 would still be dubious. Perhaps a middle course is the least objectionable: Make a new index number from a new sample of stocks every ten or twenty years, using geometric means; each time that a new series is made compute overlapping figures for a few years both from the old and from the new samples: find what part of the changes in those years is due to alterations in the list of stocks, and, finally, allow for these differences as well as may be in joining the two index numbers together. The price comparisons that could be extended in this way over long periods of time would not indeed possess the accuracy of our year-to-year figures, but they would be more trustworthy than any of the fixed-base series.

REVIEW

1. State Professor Mitchell's general conclusion.
2. Would his conclusion apply to commodity wholesale prices; to commodity retail prices designed to measure changes in the "cost of living"? Why? (Answer these questions in the light of both the *Text* discussion and the above article.)

REVIEW PROBLEMS

Index Numbers and Averages

1. Change of Base and Use of the Arithmetic Mean. Average of Relatives. (See *Text*, page 318 and note 2.)

Using the absolute price data on page 296 of the *Text* recompute a simple average of relatives price index number for each of the years with 1914 as the base. Compare the numbers as thus determined with those for 1912 and 1913 obtained by dividing the indexes for these years, as given on page 296, by the 1914 number.

What conclusions do you draw from this experiment relative to the methods of base shifting when dealing with an average of relatives index number? In what respects are the contents of note 2 on page 318 borne out?

2. Comparison of a Simple and of a Weighted Index Number Series.

Using the price data on page 296 of the *Text*, compute weighted average of relatives index numbers for 1912, 1913, 1914. Compare the weighted with the simple numbers. Arrange the data in the form of a table, properly label it and give it a correct title.

Use the following weights:

Item	*Weight*
Total	353
Corn	85
Cotton	33
Oats	15
Hay	1
Hides	6
Cattle	90
Hogs	123

3. Use of Median Index Numbers in Simple and Weighted Series. By using the commodities and prices on page 296 of the *Text*, compute a median of relatives index number for each of the years 1912, 1913, and 1914. Compare these with medians obtained when the following weights are used:

Commodities	*Weights*
Total	353
Corn	85
Cotton	33
Oats	15
Hay	1
Hides	6
Cattle	90
Hogs	123

(1) What effect do the weights seem to have?

(2) Would this be true if another system of weights were used?

(3) Would this be true if the *order* of the weights assigned to hay, oats, corn, and cattle were changed? If the remaining weights were concentrated on the commodity cotton? Would the degree of concentration be significant? Why?

(4) Which of the above questions would you answer differently in case the arithmetic mean were used? In what way?

(5) Arrange your data in the form of a table, properly label it, and give it a correct title.

4. Base Shifting and the Use of the Median. (*Text* page 322.) Using the unweighted medians of relative prices determined in Problem 3, shift the base to 1914 by dividing through by the 1914 number. Compare these results with those obtained by recomputing throughout the relatives on the 1914 base. (See data on Problem 1.)

In what ways do your results bear out the contention of the *Text*, p. 322, relative to the use of median in Index Numbers? Be specific.

CHAPTER IX

DESCRIPTION AND SUMMARIZATION — DISPERSION AND SKEWNESS

THE NATURE OF STATISTICAL KNOWLEDGE [1]

A careful consideration of the history of statistical science leads to the conclusion that statistical methods are used for two sorts of purposes, or to gain two sorts of knowledge about events or things.

A. On the one hand the statistical method finds one of its chief uses in furnishing a method (and the only one known in science) of describing a group in terms of the group's attributes, rather than in terms of the attributes of the individuals which compose the group. . . .

What sorts of positive, definite, and exact knowledge do statistics give us?

1. Precise knowledge of the *composition* of groups or masses. This is the knowledge gained by *counting*. Suppose we find a basket containing a number of balls of several different colors, and proceed to count them with the following results:

7 Reds
9 White
2 Black
1 Green

[1] Adapted with permission from Pearl, Raymond, *Modes of Research in Genetics*, Macmillan, 1915, pp. 79–100.

Such a count furnishes us at once with a great deal of perfectly definite and precise information about this group or population of balls. For example, the count tells us that it will never be possible to draw more than one pair of balls of which one member is green. This is a definite attribute of this population which may be used to differentiate it from other populations. In this particular population only one green ball occurs.

This sort of knowledge derived by counting is perfectly definite and precise so far as it relates to the particular group or mass which it concerns in any particular case. It does not involve any approximation, or probability, and is as precise as knowledge of the individual. It, however, pertains to the group. It forms a part of a proper scientific description of a group.

2. Knowledge of certain *abstract qualities* of groups or masses. This knowledge is obtained by calculation from the counted data. The more important of the abstract qualities of groups are:

a. The *center* or *typical condition* of the group; or the condition about which the individuals composing the group cluster. This is variously measured: by the arithmetic mean, which gives the center of gravity of the group, by the median, which tells the point on either side of which exactly half the individuals fall, by the mode, which tells the point of greatest frequency of occurrence in the group, etc.

b. The degree of *individual diversity* comprised in the group. This attribute, called the variability of the group, is again variously measured: by standard deviations, coefficients of variation, etc.

c. The *degree of symmetry* of the distribution of the individuals composing the group. This is measured by the skewness or other related constants. . . .

One point here we must be quite clear about. This is that the kind of knowledge discussed under this heading 2 is just as definite and precise, and involves as little approximation and indeterminism, as does any piece of individualistic knowledge, *so long as we confine our attention solely to the particular group discussed in a particular single case.* We are accustomed to stating means, for example, with probable errors. But this is only because it is proposed to extend the conclusions beyond or outside of the particular group and the particular instance for which the mean was calculated. *For that group and that instance* the mean is perfectly exact and precise to that degree of precision denoted by the unit of measure used, assuming that no arithmetical mistakes have been made in its computation. Thus suppose one measures the stature of three men to the nearest inch, and then calculates the average. The result is, without any probable error, the average height, at the particular moment when they were measured, of *those three men exact to the unit of measurement used.* It describes and measures precisely an attribute of those men considered as a group. But if we were to consider this result from the viewpoint of whether it gave a reasonable measure of the average height of men in general, or from the viewpoint of whether it gave a proper value for the mean height of these men when repeatedly measured under varying conditions, it would clearly be subject to a large probable error. It would, in point of fact, have lost its character of precise and definite knowledge, and have become a more or less poor approximation.

3. Precise knowledge of the *degree of association* or contingency between different events or characters within a group. This is furnished by the method of correlation in one or another of its various forms. By this general

method we are able to measure precisely the degree of resemblance between the individuals composing a group in respect to one or more characters. So long as attention is confined to the particular group on which the measurement is made, and to that group alone, and to a single instance (in time) the knowledge gained is precise. It is a part of the description of the attributes of that group. When we pass from that particular group to other groups, or individuals, our results are no longer precise, but inferential, and the probable errors tell us something about the degree to which the inference is trustworthy.

Summarizing the results of the above analysis, we see that the statistical method can

1. Furnish precise descriptive knowledge about groups. This knowledge is of various sorts. It is definite and precise so long as attention is confined solely to the particular group and the particular instance on which it is based.

2. The knowledge gained by the statistical method, as we have analyzed it above, precise though it may be, *pertains to the group and not to the individual.* It is exact knowledge about the composition, or attributes, or contingencies of masses or groups.

3. This ability to describe groups in terms of the groups' own attributes, which is an unique property of the statistical method, is extremely useful in the practical conduct of scientific investigations. It makes the statistical method an absolutely essential adjunct to every other scientific method, and particularly to the experimental. This fact is just now beginning to be recognized by some experimentalists and hailed as a rather original thought. It is not new.

B. We may turn now to a wholly different aspect of the statistical method, wherein it is used for the purpose of predicting or estimating the probable or the approximate condition in the *individual* from a statistical examination of the condition in the mass or the group. Resort is had to the statistical method for this purpose primarily in those cases where the outcome of the event, or the condition of the thing, is determined by the combined action of a large number of small causes, each about equally influential upon the final result.

Originally the statistical method was only employed for this second purpose in cases where, because of the multiplicity of the cause groups involved in the determination of the event, and the consequently small effect of each, it was impossible to make any reasonable prediction regarding an individual from an examination of that individual alone. Such employment might be considered legitimate, though not very fruitful, on the ground that prediction so made, uncertain and doubtful as it may be, is after all perhaps better than no prediction at all. As time has gone on, however, there has been an increasing tendency to assume that this use of the statistical method had general a priori validity and could be profitably employed in all sorts of cases. This point of view reaches, it seems to me, its limit in the following sentence from Royce. "There is, therefore, good reason to say that not the mechanical but the statistical form is the canonical form of scientific theory, and that if we knew the natural world millions of times more widely and minutely than we do, the mortality tables and the computations based upon a knowledge of averages would express our scientific knowledge about *individual* events much better than the nautical almanac would do."

This leads us to consider carefully the general question of the validity on the one hand, and the usefulness on the other hand, of this whole second mode of employment of the statistical method. It is the one which has attracted the greatest attention because of its essentially spectacular nature coupled with a sort of mysteriousness bordering upon the miraculous. It seems a wonderful, indeed almost a superhuman, accomplishment to be able to say in the manner of the oracles of old, "So many men will commit suicide next year."

Since Clerk-Maxwell introduced statistical modes of reasoning into physical science there has been an ever increasing tendency to regard the universe as organized on a statistical plan. This has come to carry with it two implications, one of which is quite fallacious and the other partly so.

The first of these is that the individual events, of which all the causes are not precisely known to us, are indeterminate. Such an assumption is of course unwarranted. Because we do not know all the causes leading to a particular event does not mean that that event is any the less precisely determined by the course of antecedent events. Consider a box containing 100 consecutively numbered cards. Suppose one card were to be drawn and that it bore the number 36. It would be quite impossible to formulate precisely all the causes which led to the drawing of the number 36 on the particular occasion considered, but it is equally impossible to conceive that this result was not definitely "caused." In other words, there clearly was a whole train of antecedent circumstances, which taken all together definitely resulted, *and could only have resulted*, in the drawing of the number 36. The too prevalent conclusion that the application of the statistical method or statistical modes of thought implies phenomenal indeterminism in the individual case is totally fallacious.

The second currently accepted implication of a statistical view of the universe is that in general a particular event or phenomenon is the outcome of the combined action of a great number of causes, each of which alone produced but a small part of the final total effect. There is clearly so much truth in this point of view as is included in the fact that individual events or phenomena do, in some degree or other, vary, and further these variations in general distribute themselves more or less in accord with the well-known laws of errors. But the assertion that events are individually the outcome of the action of great numbers of causes, each of which had a small part and a part significantly equal to that played by every other one of the causes concerned in the final result, is only true if the "universe of discourse" is indefinitely extended in time. But *practically* science works in a definitely and rather narrowly limited universe of discourse so far as concerns time. One of the causes for the writing of these lines is that a certain worthy was not shipwrecked in voyaging to this country nearly 300 years ago, since if he had been shipwrecked presumably I should not exist and therefore could not write these words. But practically this cause had very little to do with determining that I, being here in existence, should write this book rather than do various other things which I might have done instead. It undoubtedly is true that a vast number of small causes do play a part in the determination of any particular event. But, in many of the events, at least, in which science is interested, these multitudinous minor causes do not play any significant part in the differential determination of a particular event at a particular instant of time. There is in connection with the causation of most events some one or two, or at most a very few, outstanding cause groups which, for all practical purposes, at a given moment completely determine their occurrence. The total effect of all the vast number of other minor causes concerned in the remote past is so minute, as compared with the part played by the really determinative ones at the moment, as to be negligible. In other words, all natural cause groups are not small, nor of equal (balanced) values in the final determination of the event to which they relate, provided we confine ourselves to the time limits of finite practical operations. . . .

The fact that all natural causes or cause groups are not equally significant quantitatively is, of course, what makes the experimental method fruitful — one might even say possible — in science. The very essence of the experimental method is that the conditions for the happening of an event are so arranged that the influence of one putative causal factor may be tested at a time. If with a radical change in this one factor, whilst all others remain, so far as may be, constant, no change in the happening of the event is observed, the experiment has shown that this particular factor has no *significant* causal relation to the happening of the event. If a marked change in the happening of the event is observed always to follow the change of conditions of operation of the factor under investigation, then clearly this factor plays a determinative part. In other words, it is a fundamental logical prerequisite of the experimental method if it is to be successful (that is, contribute to knowledge) that it operate in a universe in which all causal factors are not of equal quantitative significance at any given instant of time.

Clearly experimental analysis of this sort would have quickly discovered, if the common sense of men had not long previously shown, that the course which a particular event is going to take is not immediately the result of the action of an indefinitely large number of individually insignificant causal factors, but that it is the outcome of the action of a few immediately *determinative* factors and the effect of the indefinitely large number of historically antecedent small causes is insignificant in the sense of being differential. Generalized, the point may be put in this way: an event A is about to happen. It may happen in any one of n different ways, each one of which ways may be designated by a letter, l, p, r, t, etc. Now an indefinitely large number of causes are concerned in bringing it about that the event A is going to happen, and that it can equally well happen either as l, p, r, t, etc. In other words, the setting of the stage for the event has involved a vast number of small and balanced causes. But the causes which are differential in the particular case, that is, which determine that A shall happen in the p way this particular time, and not in the l, the t, or any other way, are, in general:

1. Few in number.
2. Immediate in time.
3. Large in relative quantitative effect.

The point under discussion may perhaps be made plainer by a homely illustration. Suppose a man steps up behind a mule and prods the creature with his walking stick. The human intellect is unequal to the task of predicting exactly, in the particular case, what precise portion of the man's body the mule's hoof will land upon. A multitude of minor causes will affect this: The relative height of the man and the mule, the age of each, the place poked with the walking stick, the degree of fatigue of the mule, the temperature, the season of the year, and countless other things have an influence in determining just the precise spot where the mule's foot and the man's body come together. These could be investigated statistically and tables drawn up from which one could predict the part of the man which would most probably receive the hoof. But what a silly, futile piece of business this all would be, since clearly the influence of all of these small causes on what happens to the man is stupendously overshadowed by the results of two factors; namely, putting himself behind a mule and prodding the animal with a stick. Of course, a vast number of antecedent causes are involved in the setting of the stage, but these are not differential in the determination of the end event of the series.

The preceding illustration has nothing directly to do with science, but the essential point involved operates in the use of the statistical method as a weapon of scientific research. This method being, as we have seen elsewhere, only a descriptive method, it cannot, any more than any other descriptive method, tell us anything *directly* about the causes involved in the determination of any events or phenomena under consideration. It may be of great aid, in combination with the experimental method, in helping to arrive at such knowledge, but alone and of itself it cannot directly furnish knowledge of causes of individual events. Yet the statistical method, particularly in that phase of it which we have here under discussion, which essays to predict the probable condition of the individual from the knowledge of the mass, *seems* to furnish information about causes. It wears a specious air of bringing a kind of knowledge which in reality it not only never does, but from the very nature of the case never can furnish.

Let us consider now a little more in detail the nature of the prediction of the probable condition of the *individual* from a knowledge of the *mass* or group. It has been shown in an earlier section that statistics give perfectly definite and precise, and often very useful, knowledge about masses or groups. We are now, however, not concerned with this as group knowledge, but rather with one use to which such knowledge has been put. This use is that which is comprised in the subject of statistical probabilities, and which involves the drawing of conclusions as to the *probable* condition of the individual, based on an *exact* knowledge of the mass.

In order to approach the subject in the simplest way let us consider a concrete case. Suppose a problem of the following sort were to be set before us for answer: What is the probability that, at some chosen moment of time, the next birth to occur in, let us say, the city of Baltimore, will be of a white child. Now if we look at this as a question in statistical probability the appropriate way, of course, to go about solving it is to turn up the registration reports for the city of Baltimore covering a period of years, and find out what is the proportion of white to colored births in that city. Then, by the simplest theorem in the calculus of chance, the probability that the next birth will be of a white child will be given by a fraction of which the numerator is the number of white children born in Baltimore and the denominator is the total number of children born in Baltimore, both figures including the same period of time. The difference between the fraction so obtained and 1 will be the probability that the next birth will be of a child not white; that is, colored. When we have obtained such a fraction we have a definite piece of statistical knowledge, but of just what use is it so far as concerns the individual case? It implies no biological knowledge of any kind; no knowledge of the laws of heredity. It really adds essentially, it seems to me, to the sum total of the world's knowledge only one thing. That thing is the proper betting odds on what the color of the next child born in the city will be. This knowledge would really be useful, in a pragmatic sense, only provided some one wishes to gamble upon that event.

Of course the statistical count, on which the probability is based, in itself furnishes definite and precise information about the population of Baltimore, *as a population*. This may be useful. What we are now considering, though, is knowledge about individual cases.

Let us see what a totally different kind of ability to predict the future event in an individual case is gained when we take into account one single biological fact of an individualistic instead of a statistical character. Suppose, that is to say, that we are informed that the mother of the next baby to be born in Baltimore is black. It needs no argument to show how much more precise is our prediction as to the color of the next baby under these conditions.

This illustration brings out clearly the difference between the two possible bases for the prediction of a future event. On the one hand, such prediction may be based on statistical ratios. This means merely a count of an indefinitely large past experience regarding the occurrence or failure of the event, but in no way takes into account the causes which underlie the happening of the event in any particular case. On the other hand, we have the prediction which is based on a definite knowledge of the determinative causes which bring about the happening of a particular individual event of the sort in which we are interested and about which we are to predict. There can be, it would seem, no comparison between the usefulness, in the pragmatic sense, of these two kinds of knowledge. The statistical knowledge on which a statistical prediction is made is essentially the most sterile kind of knowledge that one can possibly have so far as concerns the individual event. It merely gives one the betting odds for or against the occurrence of an event, and absolutely nothing more. Now a wager, however large, in the scientific sense neither discovers, expounds, nor is a criterion of the truth. Bets, in other words, are not evidence, though the statistician sometimes seems to forget this, and to deal with statistical ratios as though they had probative worth in regard to phenomena.

On the other hand, a prediction based on experimentally acquired knowledge of the determinative cause of the individual event brings with it a real knowledge of a natural phenomenon. The predictions so made may not always turn out correct, but when they do not, it incites us to investigate the particular disturbing factor which under such circumstances may overwhelm the normally determinative cause of a particular event.

. . . If, as has been suggested, that part of the statistical method which uses the calculus of probability as a basis for the prediction of future events gives only a knowledge of betting odds, one may ask: what about the whole concept of probable error? The value of this concept in scientific research is unquestioned. Yet plainly the whole concept has its basis in the calculus of probability. Has not our discussion led us unwittingly into a serious contradiction?

I think not. Let us examine the probable error concept a little more carefully than we have yet done. Suppose we read that the mean length of the thorax of a thousand fiddler crabs is $30.14 \pm .02$ mm. Just what does this actually mean? Accepting the figures at their face value, or, put another way, assuming that the mathematical theory on which the probable error was calculated was the correct one, the figures mean something like this: If one were to take, quite at random, successive samples of 1000 each from the total population of fiddler crabs and determine the mean thoracic length from each sample, these means would all be different from each other by varying amounts. In other words, no single sample would give us the absolutely true value of the mean thoracic length of the whole fiddler crab *population.* This true value is in an absolute sense unknowable, because, for one reason, always we must come at the finding of it by the way of random sampling, and sampling means variation. Now it is an observed fact of experience that the variations due to random sampling distribute themselves according to a definite law of mathematical probability. Knowing this law, it is clearly possible to state the mathematical probability for (or against) any particular deviation or variation occurring as the result of random sampling. Exactly this is what the probable error does. It says, in the particular case here considered, that it is an even chance, that a deviation or variation in the value of the mean as great as or greater than .02 mm. above or below will occur as a result of random sampling. Or, put in another way, if we took successive samples of 1000 each from this crab population, it is an even bet that the value of the mean from any sample would fall between $30.14+.02=30.16$, and $30.14-.02=30.12$.

Now all the knowledge that this probable error furnishes is this: that if a man were to say, "I'll bet a thousand dollars that the mean thoracic length of the next sample of fiddler crabs you measure will be either over 30.16 mm. or under 30.12 mm.," one would not be justified in offering odds. He could wager on even terms. Either party involved in the transaction would be as likely to lose (or to win) as the other.

Putting the case in this way, it is clear that this is the same kind of knowledge which comes from an examination of probable errors as that discussed in the preceding section. It is a knowledge of betting odds. It has no necessary relation *per se* to any physical, chemical, or biological laws. It merely informs one how he may safely gamble on an event if he is so minded and can find some one else ready to do the same thing.

Wherein lies the value of the probable error concept for science, then? Simply in that it serves as a test or check on every mode of research in science. So far as I can see, the calculus of probability, in and of itself alone, is not and never can be an effective weapon of research for the discovery of truth in phenomenal science, be it physical or biological. Yet it operates as an ever-present test of the trustworthiness of the results obtained by modes of research which are in themselves adapted to making discoveries about phenomena. The student of probability says something like this to the experimentalist: "Yours is the way to find out the significant underlying causes of phenomena. Let it be practiced with all zeal, but let it be remembered that you operate in a finite way in a finite universe, and consequently all your results are subject to such fluctuations and variations as experience has shown arise from random sampling. I regret that I cannot directly and alone discover significant causes, but at any rate I can furnish you a test whereby you may reasonably judge whether your result is significantly influenced by these fluctuations of random sampling."

To sum the whole matter up: I have tried to show that the statistical method in science has been used to do two things.

The first of these is a unique function of the method —

to furnish a description of a group of objects or events in terms of the group's attributes rather than those of the individuals composing the group. Herein lies the great value of the statistical method. It is, however, a descriptive method only and has the limitations as a weapon of research which that fact implies.

The second purpose that the statistical method has been called upon to accomplish is the prediction of the individual case from a precise knowledge of the group or mass. This involves something really additional to the statistical method *per se;* namely, the mathematical theory of probability. We have seen that this side of the statistical method gives only a somewhat sterile kind of knowledge so far as concerns individuals; namely, a knowledge of betting odds. The theory of probability grew up about the gaming table, not in the laboratory. Its place in the methodology of science is not an independent one. By it alone one cannot discover new truths about phenomena. But it is a highly important adjunct to other modes of research.

Plainly, however, one cannot regard statistical knowledge in general as a higher kind of knowledge than that derived in other ways. Nor is the statistical method to become the dominant or exclusive method of science, though it will always be useful, and in many fields an essential method. It will find its chief usefulness, first in its sphere of furnishing shorthand descriptions of groups, and second in furnishing a test of the probable reliability of conclusions.

REVIEW

1. What are the two sorts of knowledge about things or events which statistical methods help to secure?

2. In what sense may the arithmetic mean be said to be "precise and exact" — that is, a reality in somewhat the same sense as is the mode? In what sense does it become very inexact and unprecise?

3. What assumptions are made when, from the use of statistical methods, an attempt is made to predict "the probable condition of the *individual* from a knowledge of the mass or group"?

4. How does Dr. Pearl illustrate the problem of the condition of the individual from a knowledge of the group in re likelihood of the birth of a black child in Baltimore?

5. What is the probable error and what is its function?

The Horizontal Zero in Frequency Diagrams [1]

It is a generally accepted rule of graphic presentation that a zero, used in a diagram as a point of reference, should be included in the diagram. This rule, while it is observed in most statistical work, is almost universally disregarded in the drafting of frequency diagrams.

Diagram 1, presented herewith, is a frequency graph of a common type, based on the weights of 738 men.[2] Weights are indicated on the base line, and the per cent of cases corresponding to any given weight is proportionate to the vertical distance from the base line to the curve. A zero line is the most conspicuous feature of this diagram, but inspection of the figure shows that the presentation implies two zeros, and that only one of these is shown. The vertical scale, representing percentages, begins at the zero base line, but the horizontal scale, representing weights, begins at 90 pounds. It is the purpose of this paper to state reasons for including the horizontal zero, to direct

[1] Adapted with permission from Clark, Earle, "The Horizontal Zero in Frequency Diagrams," *Quarterly Publications of the American Statistical Association*, June, 1917, pp. 662–669.

[2] The data are for 738 men born in Wales, as shown in Yule's "Introduction to the Theory of Statistics," p. 95. For convenience in presentation, the extremes of the distribution have been arbitrarily shortened.

attention to a type of frequency diagrams to which these reasons do not apply, and to illustrate methods of drafting.

A frequency diagram is plotted for the purpose of showing the significant facts about a series of variables. The graphic form is used rather than a frequency table or text statement because most people, even most statisticians, find it easier to perceive and appreciate these significant

DIAGRAM 1.—WEIGHTS OF 738 MEN, SHOWN WITHOUT HORIZONTAL ZERO

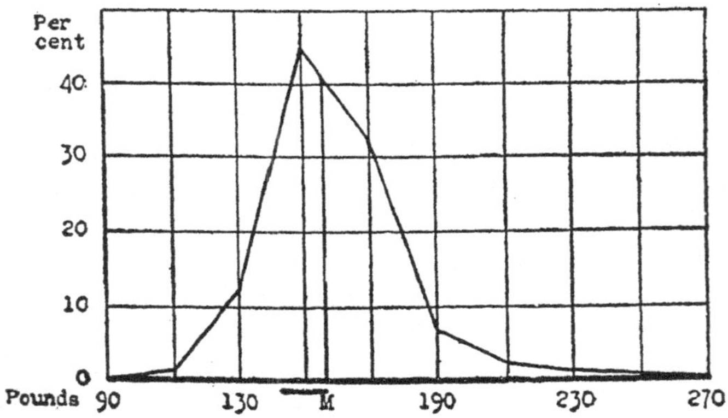

facts by looking at a diagram than by studying a column of figures. The essential facts about a variable series are: (1) the mean, median, or other measure of central tendency, and (2) the distribution of the values about this central tendency. These facts are interdependent. It is a simple matter to compute medians or means, but these measures do not reveal the whole truth about a distribution; they may be seriously misleading unless shown in relation to the distribution of the individual values.

On the other hand, the distribution is not in itself significant unless related to the central tendency. Stated in pounds and ounces, the average deviation of the weights of a group of 1000 elephants would doubtless be far greater than the average deviation of the weights of 1000 canary birds, but this would not necessarily mean that the weights of elephants are relatively more variable than the weights of canary birds. In order to determine the true variability of a series it is necessary to relate the measure of dispersion to the measure of central tendency. This may be done by computing a coefficient of dispersion — a ratio which expresses the dispersion as a proportion of the measure of central tendency.

It follows that, if a frequency diagram is to serve the purpose for which it is intended, it must show, with all possible clearness and effectiveness, the distribution of the individual values, the central tendency, and the relation of the distribution to the central tendency. Diagram 1 shows the distribution of the measures. Does it also show, with the emphasis required, the two other essential facts?

On Diagram 1 the median is indicated in the usual way — by a vertical line dividing into two equal parts the surface of the figure inclosed by the curve and the base line. This line is sometimes referred to as the median line, but the designation does violence to the principles of graphic presentation. In diagrams, lines or areas are, or should be, proportionate to the quantities they represent. The length of the so-called "median line" is not proportionate to the median weight of men; it is proportionate rather, as the class interval for the distribution is 20 pounds, to the approximate number of men whose weights fall within limits fixed, respectively, at 10 pounds below and at 10 pounds above the median weight. The line represents, in other words, not the median value for the series, but a number of cases. There is nowhere on the diagram a line representing by its length, or a surface representing by its area, the median weights of the men.

The median can be determined, it is true, by referring to the scale at the foot of the figure. As the point of intersection of the so-called "median line" with the base line falls at 156 pounds, as indicated by the horizontal scale, it follows that this value is the median, but the result is not

Diagram 2.—Weights of 738 Men, Shown with Horizontal Zero

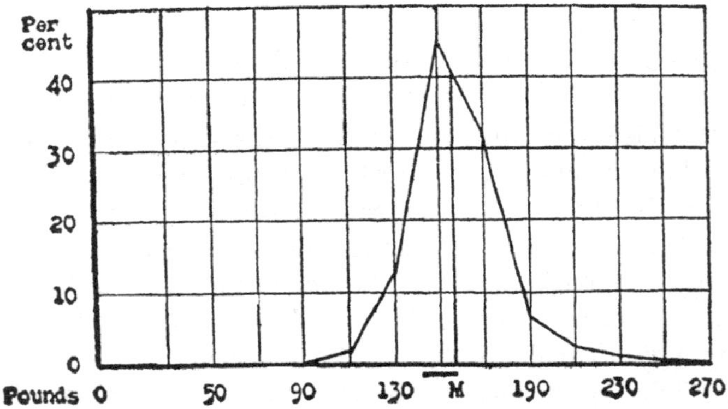

obtained by the graphic method. The figures on the scale are not graphic representations any more than are the figures of a table or a text statement.

The median can, however, be shown by the graphic method by so extending the base line that the horizontal scale will include the zero. This method has been followed in preparing Diagram 2. In Diagram 2 the horizontal distance from the vertical line at the left of the figure to the so-called "median line," measured on the base line or along any abscissa, represents the median weight of the men.

If the inclusion of the horizontal zero is required for a complete graphical representation of the median, it is even more essential as a means of showing the relationship of the dispersion to the median. As Diagram 1 contains no graphical representation of the central tendency, it follows that it affords no graphical representation of the relation between the central tendency and the dispersion. The dispersion of the series is indicated by the form of the curve and also by a line beneath the base line, proportionate in length to the average deviation (14.2 pounds), drawn to scale and extending to the left of the median. By including this line, the dispersion is reduced to a single graphical expression, but the diagram contains no graphical representation of the median with which either the line or the curve can be compared.

An effective graphical representation of the relationship between the central tendency and the distribution is found in Diagram 2, in which the median, represented by the distance between the horizontal zero and the vertical "median line," can be compared both with the surface of frequency, as indicated by the curve, and with the line representing the average deviation. The ratio of the length of this line to the distance from the horizontal zero to the median line is equivalent to the coefficient of dispersion.

The difficulties arising from the omission of the horizontal zero are further illustrated in Diagram 3, in which the weights of the 738 men are compared with the weights of 279 thirteen and fourteen-year-old school boys.[1]

In Diagram 3 the scales for pounds are identical in both figures. The appearance of the diagram suggests that

[1] The data, which are for boys attending the Worcester, Mass., public schools, are from a report by Franz Boas and Clark Wissler, published in the report of the U. S. Commissioner of Education for 1904.

the two distributions are very much alike; as the figure for men has a greater spread at the base line than that

DIAGRAM 3. — WEIGHTS OF 738 MEN AND 279 BOYS, SHOWN WITHOUT HORIZONTAL ZEROS

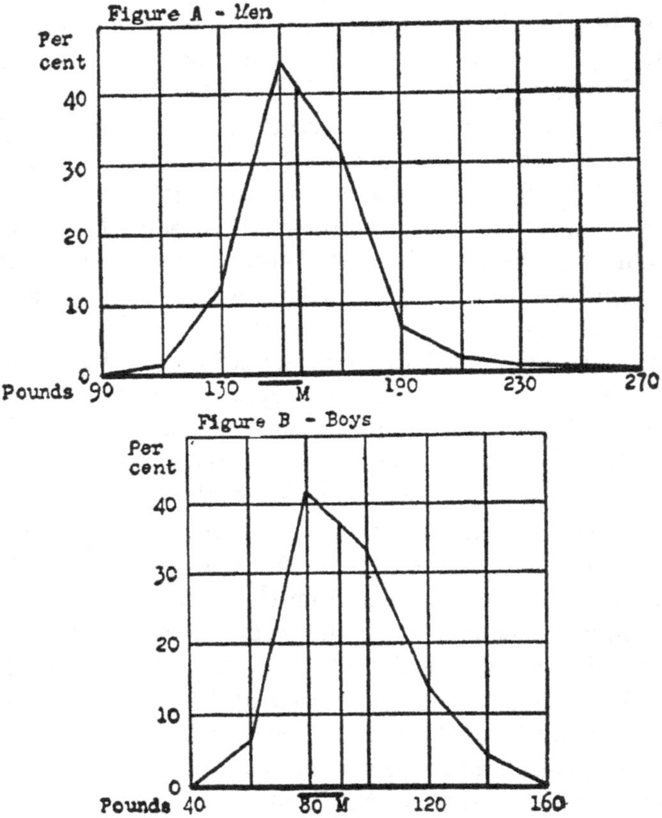

for boys it would seem that the former represents, if anything, the wider dispersion. This impression is not borne out by the data. The actual dispersion (average deviation) is, roughly, the same for the two series: 14.2 pounds for the men and 14.3 pounds for the boys. But as the median for the men is 156.3 pounds, and that for the boys 90.8 pounds, computation shows that the significant measure of relative variability, the coefficient of dispersion, is .157 for the boys and only .091 for the men. In other words, the dispersion of the weights of the boys is 15.7 per cent of the median weight of boys, while for the men the dispersion of the weights is but 9.1 per cent of the median weight of men. The apparent similarity of the two distributions represented in Diagram 3 is, therefore, accidental and the diagram is misleading.

It may be said that any one using Diagram 3 could determine the relative dispersions by a study of the figures of the scales; that the scales show the medians, and that it is not impossible to relate these medians to the dispersions. This is true, but, as the same facts can be determined from a frequency table, the argument offered is merely an argument for not using graphical representations for comparing two or more series of variables.

Diagram 4 shows in graphic terms the true relationship between the dispersions. The base lines of Figures A and B of this diagram have been carried out to zero, and the scales have been so adjusted that the distance from zero to the median is the same in both figures. It is now possible to view the dispersions in their relationship to the central tendencies. The lines representing the average deviations, as well as the contours of the curves, show very clearly that the weights of boys are much more widely dispersed than the weights of men.

The fact that in Diagram 4 the surface inclosed by the curve and base line of Figure B is much greater than that inclosed by the curve and base line of Figure A might lead an incautious observer to assume that the dissimilarity in the appearance of the figures is due to a difference in the

DIAGRAM 4.—WEIGHTS OF 738 MEN AND 279 BOYS, SHOWN WITH HORIZONTAL ZEROS

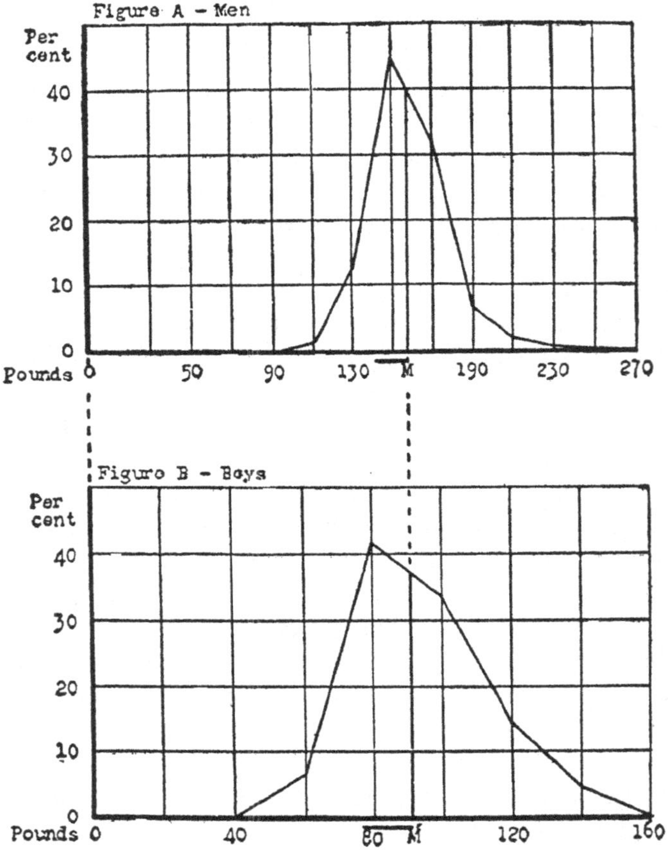

number of observations — that the number of boys exceeds the number of men. Such an inference would be unwarranted. As numbers have been reduced to percentages, 100 per cent is the total for each group. The values are plotted upon the ordinates; hence, the spaces between the ordinates, and the areas inclosed by the curves and the base lines, are without significance. It is believed that the diagram affords a correct interpretation of the data; that it gives an impression of two groups of which one is somewhat closely clustered about its central tendency, while the other is much more widely dispersed.

It should be noted that there is an important group of frequency diagrams to which the arguments in favor of including the horizontal zero, which have been stated in the preceding pages, do not apply. These are diagrams of distributions in which the zero cannot be exactly located. In the so-called normal frequency distribution the base line and the ends of the curve are in asymptote — the ends and the base line are tangent at infinity. It follows that, in plotting probabilities, or results in the psychological field which are based not upon concrete measurements but upon rankings, the horizontal zero cannot be shown.

But it is also impossible to show a zero based upon data of this kind in any type of diagram, and this is true whether the zero is vertical or horizontal. If the horizontal zero cannot be shown in a frequency diagram representing the distribution of schoolboys with reference to a given mental trait, as determined by the rankings of competent judges, neither can a zero be shown in a diagram in which the ability of any one of these boys at successive tests is indicated by a historical curve. It is possible to present a horizontal zero in a frequency diagram for any data for which a vertical zero for an ogive curve can be shown.

A practical objection to the inclusion of the horizontal zero is the fact that additional space is required. But this objection is no more applicable to the horizontal zero in frequency diagrams than to the vertical zero in line diagrams. The inclusion of the vertical zero in diagrams of the latter type is the established practice. And an inspection of the diagrams presented with this paper makes it clear that the inclusion of the horizontal zero presents no serious difficulties. A case will occasionally be encountered in which the dispersion constitutes so small a proportion of the central tendency that the zero, whether horizontal or vertical, must be omitted, but such cases are most exceptional.

The arguments and the illustrations presented in the preceding pages seem to support the following conclusions: In frequency diagrams, where the position of the horizontal zero is exactly ascertainable, and where the dispersion is not too small in proportion to the measure of central tendency, the horizontal zero should be included in the diagram. This means that the horizontal zero should be included in a frequency diagram in all cases in which a zero for similar data would be included in any type of diagram. Without the horizontal zero the frequency diagram does not afford a complete graphical representation of the central tendency nor of the relationship of the central tendency to the distribution.

REVIEW PROBLEMS

Dispersion and Skewness

1. Dispersion.

(1) Using the data in Chapter VIII, pp. 348, for expenditures for breakfast, dinner, and supper, express both absolutely and relatively the dispersion in different expenditure series by the cumulative or moving-range method. Put your data in the form of a single table. (See *Text*, page 383.) Reduce the measures of dispersion to coefficients: Relatively how do the series compare?

(2) Average Deviation.

Using the data as in (1) above, compute the average deviation by the short-cut method. Arrange the data in the form of tables. (See *Text*, pages 396–398.) Test your result by computing the average deviation from the true average. Reduce your measures of dispersion, based upon the average deviations, to coefficients. Relatively how do the distributions stand?

(3) Using the data as in (1) above, compute the standard deviations. Arrange your data in the form of tables. (See *Text*, pages 404–405.) Compare the standard and average deviations. Do the contentions in the *Text*, pages 402–403 and 406, seem to be borne out? Reduce the measures of dispersion based upon the standard deviation to coefficients. Relatively how do the distributions stand?

(4) Quartile Deviation.

Using the data as in (1) above, compute both the quartile measures and coefficients of dispersion. Compare the quartile measures and coefficients of dispersion with those based on the standard and average deviations. Arrange your comparison in the form of tables. Do the contentions respecting the quartiles, found on pages 408–409 of the *Text*, seem to be borne out?

2. Skewness.

Using the data in (1) above, compute the quartile measures and coefficients of skewness, and the coefficients based upon the standard deviations. Is the rule on page 417 of the *Text* respecting the positions of averages borne out in these cases? What variations are there from the ideal?

3. Dispersion and Skewness.

Formulate a general statement summarizing the functions and merits in statistical analysis of measures and coefficients of dispersion and skewness. Illustrate the points made by referring to your results in the above problems. Revise your answer to Problem III on *Tabulation* in the light of your measures and coefficients of dispersion and skewness.

CHAPTER X

COMPARISON — CORRELATION

The Limits of Statistics [1]

. . . It is, however, a fact too well recognized to require specific illustration that statistics, on its objective and mathematical side, presents at best but a rearrangement of the data. The data, thus marshaled, cannot in themselves provide a solution to any social problem: they merely constitute a problem. In fact, the most signal merit of statistics consists perhaps in the very aptitude of that method to bring to the surface problems which otherwise might never be recognized. But the solution of such problems can only be reached within the level to which the data themselves belong, and thus falls to the lot of the sciences representing the conceptualizations of the particular set of data, whether this be biology, or psychology, or sociology. There is thus good common sense in the popular saying that statistics can be made to prove anything, implying that it is the interpretation of the statistical material which counts, and that, if the interpretation is arbitrary, the mathematical garb of the data is no guarantee of truth.

[1] Adapted with permission from Goldenweiser, A. A., "History, Psychology and Culture," in *Journal of Philosophy, Psychology and Scientific Methods*, October 10, 1918, pp. 567–568.

Difficulties in International Statistical Comparisons [1]

. . . The various kinds of difficulties may be broadly classified as those due to:

(1) Inadequate definition;

(2) Non-identity of definition;

(3) Absence of information showing in what particulars unlike definitions really differ;

(4) Differences in the periods of time for which statistical returns are collected. This is really a special case of differences in definition, but it is important enough to deserve special mention;

(5) Differences in the classification of statistics — another special and important case of differences in definitions;

(6) Varying degrees of incompleteness of statistics covering the same subject-matter. This case has an extensive aspect, where the statistics, though complete so far as they go, do not cover the whole ground. . . . There is also an intensive aspect, where the statistics, though nominally covering the whole ground, are incomplete through faulty collection. . . .

(7) Lack of particular kinds of information necessary to a complete comparison; and

(8) Absolute incomparability, arising from what may be called organic differences in the subject-matter, as distinct from the deficiencies in the statistics relating to that subject-matter.

[1] Adapted with permission from Weber, Augustus D., "Notes on Some Difficulties Met with in International Statistical Comparisons," in *Journal of the Royal Statistical Society*, Vol. 73, 1910, pp. 10–11.

Difficulties in International Comparison of Wages [1]

A class of statistics . . . presenting some of the greatest difficulties in comparisons, and yet one with respect to which comparisons are frequently made, is the class of wages statistics. Here it is a case of definition in the widest sense. What are wages? From current popular literature one might suppose they were a rate of money per hour, or per day, or per week, with no suggestion that such a rate may be a "standard rate," or the arithmetical average of a number of rates actually paid, or the "modal" rate actually paid, or the rate in a particular locality, or any one of a number of such things. It may happen that the only rates published are, for a certain trade in one country, actual earnings, and, in another country, the standard rates. . . . How are these to be compared without knowing the relation of actual earnings to standard rates in one country or the other? But the money rate per unit of time or work, whether standard or any other rate, is after all the least important thing about wages. If the French artisan earning *8d.* per hour is as strong and healthy, as well fed, clothed, and housed, if, in a word, he has his economic wants as satisfactorily met as the English artisan getting *10d.* an hour, can it be really maintained that economically the Frenchman is more badly paid or is worse off than the Englishman? Wages, in fact, from the international, if from no other, point of view, are not money rates, but economic goods, tangible and otherwise, which the worker can and does get in return for his labor, and wages in different countries can only be properly compared when expressed in terms of economic goods, and allowance made for the

[1] Adapted with permission from Weber, Augustus D., "Notes on Some Difficulties Met with in International Statistical Comparisons," in *Journal of the Royal Statistical Society*, Vol. 73, 1910, pp. 17–19.

different marginal values which the same goods may possess to different individuals or at least to different communities. It is, of course, well known that wages statistics are not and, in the present state of our knowledge, cannot be expressed in this way. An approximation to it is, however, afforded by the method of correcting money wages by, or rather interpreting them in the light of, what is called the cost of living. Statistics of the cost of living of particular classes in certain countries are growing in volume, though they are still too inadequate to permit of anything like an exact interpretation and comparison of money wages in terms of "real" wages. The most important recent contribution to these statistics are, so far as I am aware, the reports by our Board of Trade on cost of living in British, French, and German towns, while the United States Labor Department at Washington has issued valuable reports on cost of living in the States. From the Board of Trade reports referred to we find, *e.g.* that while money wages in England, France, and Germany may be in the proportion of 100 : 75 : 83, such wages when interpreted in the light of the cost of fuel, rent, and food in the respective countries, may be found to be in the ratio of 100 : 67 : 71. These figures may be but very rough approximations to the true level of "real" wages in the countries compared, but if the data on which they are based are fairly extensive or form a good sample from which to estimate the cost of living, they are much better than the level of money wages, and it is to be desired that authentic and detailed information on cost of living in all civilized countries may be collected and published.

But even with such additional information, the correct comparison of international wages statistics is impossible without a knowledge of the amount of unemployment experienced in different occupations in different countries. This knowledge is at present not obtained. The Trade Union unemployment figures published by the Board of Trade may reasonably be challenged, as they often are, as not affording an entirely complete statement of the amount of unemployment in this country. But such as they are, there are, I believe, no similarly extensive statistics in any other country comparable with them. The importance of unemployment as a social fact is undeniable, and every effort should be made to ascertain its real extent. This may be largely, if not wholly, accomplished by means of Trade Unions, Labor Exchanges, and Unemployment and other social insurance schemes. Until this information is forthcoming, it appears clear that wages statistics will not be capable of complete interpretation or of precise comparison.

The Coefficient of Correlation[1]

In many studies it is necessary or at least desirable to test the existence of concomitant variation between two series of variable quantities. A comparison of the plotted variables furnishes a rough, but for some purposes adequate, means of examining the relationship. Figure 1 is an example of this sort of comparison. However, the use of curves is not to be recommended for careful work because of the difficulty in selecting the proper scales and the dangers resulting from personal bias. The usual tabular method is slightly more refined but tables involve too many figures to give an adequate idea of the conditions and give no concise measure of the degree of relationship.

[1] Adapted with permission from Reed, William Gardner, "The Coefficient of Correlation," *Quarterly Publications of the American Statistical Association*, June, 1917, pp. 670–684.

The English biometricians have perfected a method of stating the degree of relationship, which was invented by Bravais about 1845. "Correlation may be briefly defined as the tendency towards concomitant variation and the so-called correlation coefficient is simply a measure of such tendency,

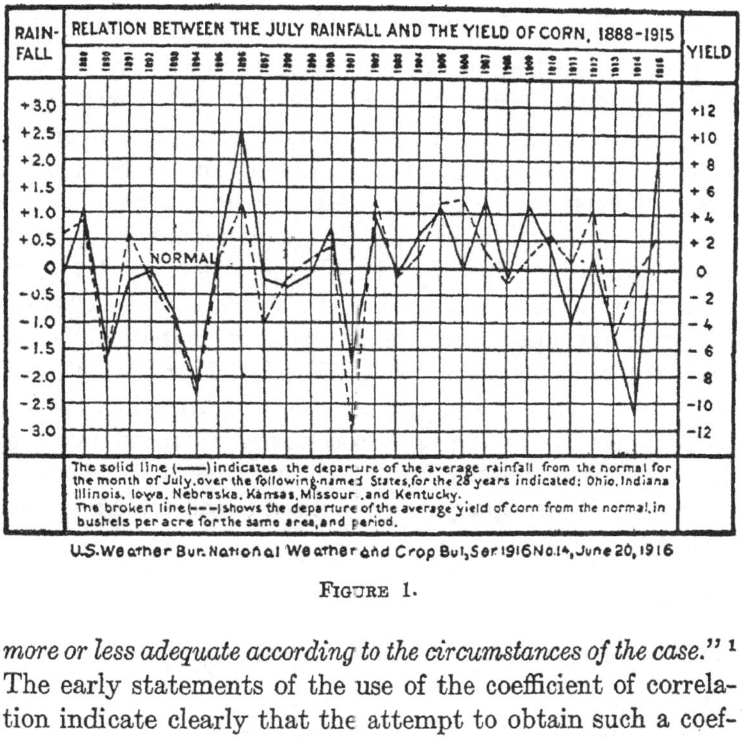

U.S. Weather Bur. National Weather and Crop Bul, Ser. 1916 No. 14, June 20, 1916

FIGURE 1.

more or less adequate according to the circumstances of the case."[1] The early statements of the use of the coefficient of correlation indicate clearly that the attempt to obtain such a coefficient from miscellaneous material is an abuse of this method of measuring relationship.[2] The material in hand should be

[1] Brown, W.: The Essentials of Mental Measurement, Cambridge, University Press, 1911, p. 42. (Italics are the present writer's.)

[2] Yule, G. U.: Introduction to the Theory of Statistics, ed. 2, London, Griffin & Co., 1912, pp. 169, 177.

investigated carefully before any attempt is made to determine the relationship by the use of the coefficient of correlation. This investigation may take the form of a correlation table or of a "dot chart" after Galton's graphic method of correlation.[1]

Method of Procedure

If the coefficient of correlation is to have any definite meaning, the procedure must be somewhat as follows:

1. The material (*e. g.* Table I) should be arranged in groups in the form of a correlation table (Table II), or, better, plotted as a dot chart (Figure 2). The table or chart should then be carefully examined to see whether the points may be generalized to a straight line, that is, whether there is a tendency for a high value of one variable to be associated with high values of the other variable and proportionately higher or lower values of the one to be associated with similar values of the other. This shows positive linear correlation. When lower values of the one are associated with higher values of the other, the correlation is said to be negative. For example, the dots in Figure 2 may be generalized to the line AB as well as to any curve.

$M'_x = 4.0$ inches

$M_x = 4.0 + \frac{3.9}{60} = 4.1$

$\Sigma x = +3.9$

$\Sigma x^2 = 112.67$

$$\sigma_x = \sqrt{\frac{\Sigma x^2}{n} - \left(\frac{\Sigma x}{n}\right)^2}$$

$M'_y = 35$ bu.

$M_y = 35 - \frac{26}{60} = 34.6$

$\Sigma y = -26$

$\Sigma y^2 = 1258$

$$\sigma_y = \sqrt{\frac{\Sigma y^2}{n} - \left(\frac{\Sigma y}{n}\right)^2}$$

[1] See Davenport, C. B., "Statistical Methods," ed. 3, New York, Wiley, 1914, pp. 42–47.

Table I

Correlation of July Rainfall and the Yield of Corn in Ohio
(Smith, J. W.: The Effect of Weather upon the Yield of Corn in Ohio. Washington, Mo., Weather Rev., Vol. 42, 1914, p. 80.)

July Rainfall				Yield of Corn			
Year	Amount	x	x^2	Bushels per Acre	y	y^2	xy
1854 . . .	2.6	− 1.4	1.96	26.0	− 9	81	+ 12.6
1855 . . .	5.8	+ 1.8	3.24	39.7	+ 5	25	+ 9.0
1856 . . .	2.6	− 1.4	1.96	27.7	− 7	49	+ 9.8
1857 . . .	4.9	+ .9	.81	36.6	+ 2	4	+ 1.8
1858 . . .	4.7	+ .7	.49	27.7	− 7	49	− 4.9
1859 . . .	1.6	− 2.4	5.76	29.5	− 5	25	+ 12.0
1860 . . .	5.8	+ 1.8	3.24	38.2	+ 3	9	+ 5.4
1861 . . .	3.3	− .7	.49	33.5	− 1	1	+ .7
1862 . . .	3.6	− .4	.16	30.0	− 5	25	+ 2.0
1863 . . .	2.6	− 1.4	1.96	27.0	− 8	64	+ 11.2
1864 . . .	2.1	− 1.9	3.61	27.0	− 8	64	+ 15.2
1865 . . .	5.7	+ 1.7	2.89	35.0	0	....	......
1866 . . .	5.1	+ 1.1	1.21	36.5	+ 2	4	+ 2.2
1867 . . .	3.2	− .8	.64	29.8	− 5	25	+ 4.0
1868 . . .	2.7	− 1.3	1.69	34.4	− 1	1	+ 1.3
1869 . . .	4.8	+ .8	.64	28.4	− 7	49	− 5.6
1870 . . .	4.7	+ .7	.49	37.5	+ 3	9	+ 2.1
1871 . . .	3.7	− .3	.09	36.7	+ 2	4	− .6
1872 . . .	6.7	+ 2.7	7.29	40.9	+ 6	36	+ 16.2
1873 . . .	6.2	+ 2.2	4.84	35.1	0	....	......
1874 . . .	3.8	− .2	.04	39.2	+ 4	16	− .8
1875 . . .	6.9	+ 2.9	8.41	34.2	− 1	1	− 2.9
1876 . . .	6.4	+ 2.4	5.76	36.9	+ 2	4	+ 4.8
1877 . . .	3.7	− .3	.09	32.5	− 2	4	+ .6
1878 . . .	5.4	+ 1.4	1.96	37.8	+ 3	9	+ 4.2
1879 . . .	4.2	+ .2	.04	34.3	− 1	1	− .2
1880 . . .	4.2	+ .2	.04	38.9	+ 4	16	+ .8
1881 . . .	3.6	− .4	.16	31.0	− 4	16	+ 1.6
1882 . . .	3.2	− .8	.64	34.0	− 1	1	+ .8
1883 . . .	4.2	+ .2	.04	24.2	− 11	121	− 2.2
1884 . . .	3.8	− .2	.04	33.3	− 2	4	+ .4
1885 . . .	3.2	− .8	.64	36.8	+ 2	4	− 1.6
1886 . . .	2.9	− 1.1	1.21	33.5	− 1	1	+ 1.1
1887 . . .	2.2	− 1.8	3.24	30.5	− 4	16	+ 7.2
1888 . . .	4.4	+ .4	.16	38.9	+ 4	16	+ 1.6
1889 . . .	4.2	+ .2	.04	32.3	− 3	9	− .6
1890 . . .	2.0	− 2.0	4.00	24.6	− 10	100	+ 20.0
1891 . . .	3.8	− .2	.04	35.6	+ 1	1	− .2
1892 . . .	3.8	− .2	.04	33.3	− 2	4	+ .4
1893 . . .	2.5	− 1.5	2.25	29.1	− 6	36	+ 9.0
1894 . . .	1.6	− 2.4	5.76	32.6	− 2	4	+ 4.8
1895 . . .	2.0	− 2.0	4.00	33.7	− 1	1	+ 2.0
1896 . . .	8.1	+ 4.1	16.81	41.7	+ 7	49	+ 28.7
1897 . . .	4.6	+ .6	.36	34.3	− 1	1	− .6
1898 . . .	4.0	.0	....	37.4	+ 2	4	......
1899 . . .	4.2	+ .2	.04	38.1	+ 3	9	+ .6
1900 . . .	4.6	+ .6	.36	42.6	+ 8	64	+ 4.8
1901 . . .	2.7	− 1.3	1.69	30.0	− 5	25	+ 6.5
1902 . . .	4.7	+ .7	.49	38.8	+ 4	16	+ 2.8
1903 . . .	3.7	− .3	.09	31.5	− 3	9	+ .9
1904 . . .	4.1	+ .1	.01	32.8	− 2	4	− .2
1905 . . .	3.9	− .1	.01	37.9	+ 3	9	− .3
1906 . . .	5.1	− 1.1	1.21	42.2	+ 7	49	+ 7.7
1907 . . .	5.4	+ 1.4	1.96	34.8	0	....	......
1908 . . .	4.1	+ .1	.01	36.1	+ 1	1	+ .1
1909 . . .	3.8	− .2	.04	38.7	+ 4	16	− .8
1910 . . .	3.2	− .8	.64	36.6	+ 2	4	− 1.6
1911 . . .	2.4	− 1.6	2.56	38.6	+ 4	16	− 6.4
1912 . . .	5.7	+ 1.7	2.89	42.8	+ 8	64	+ 13.6
1913 . . .	5.2	+ 1.2	1.44	37.8	+ 3	9	+ 3.6
		−30.2	112.67		−125	1258	+201.4
		+34.1			+ 99		
		−3.9			− 26		

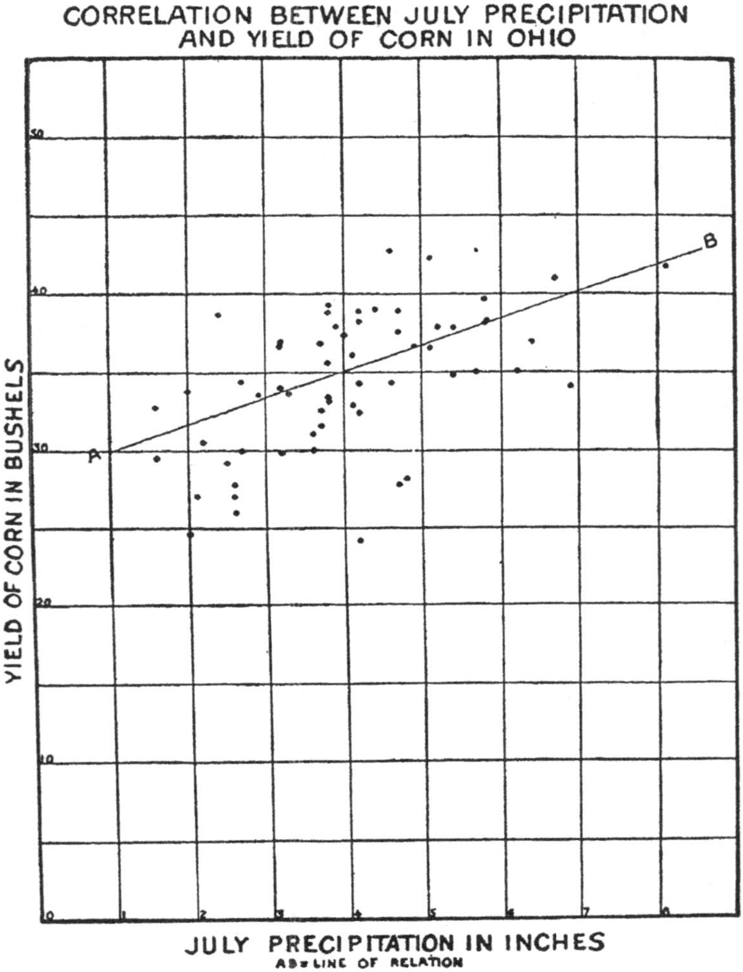
CORRELATION BETWEEN JULY PRECIPITATION AND YIELD OF CORN IN OHIO
YIELD OF CORN IN BUSHELS
JULY PRECIPITATION IN INCHES
AB = LINE OF RELATION
A
B
0
1
2
3
4
5
6
7
8
10
20
30
40
50

FIGURE 2.

$$=\sqrt{\frac{112.67}{60}-.0036} \qquad =\sqrt{\frac{1258}{60}-.2}$$

$$=1.4 \qquad = 4.6$$

$$\Sigma xy = 201.4$$

$$r=\frac{\frac{\Sigma xy}{n}-\left(\frac{\Sigma x}{n}\right)\left(\frac{\Sigma y}{n}\right)}{\sigma_x\sigma_y}$$

$$=\frac{\frac{201.4}{60}-\frac{3.9}{60}\times\frac{-26}{60}}{1.4\times 4.6}$$

$$=\frac{3.36+.03}{6.44}$$

$$=0.526\pm E_r$$

$$E_r=\pm.674\frac{1-r^2}{\sqrt{n}}$$

$$=\pm.674\frac{.723}{7.7}$$

$$=\pm.063$$

$$r=+0.526\pm.063$$

NOTE: r is not the same here as in the original paper because a single average yield of corn has been used for simplicity.

Explanation of Symbols

n number of observations (years of record).
M_x true mean July precipitation.
M'_x some arbitrary number near M_x.
M_y true mean yield of corn.
M'_y some arbitrary number near M_y.
x departure of each July precipitation from M'_x.
y departure of yield of corn in each year from M'_y.
Σx algebraic sum of departures of July precipitation.

Σy algebraic sum of departures of yield of corn.

Σx^2 algebraic sum of squares of departures of July precipitation.

Σy^2 algebraic sum of squares of departures of yield of corn.

Σxy algebraic sum of products of departures (x and y).

σ_x standard deviation of July precipitation

$$\sigma_x = \sqrt{\frac{\Sigma x^2}{n} - \left(\frac{\Sigma x}{n}\right)^2}$$

σ_y standard deviation of yield of corn.

$$\sigma_y = \sqrt{\frac{\Sigma y^2}{n} - \left(\frac{\Sigma y}{n}\right)^2}$$

r coefficient of correlation.

$$r = \frac{\frac{\Sigma xy}{n} - \left(\frac{\Sigma x}{n}\right)\left(\frac{\Sigma y}{n}\right)}{\sigma_x \sigma_y}$$

E_r probable error of the coefficient of correlation.

$$E_r = \pm .674 \frac{1 - r^2}{\sqrt{n}}$$

TABLE II. CORRELATION TABLES SHOWING THE RELATION BETWEEN JULY PRECIPITATION AND THE YIELD OF CORN IN OHIO

(From Smith, J. W., The Effect of Weather on the Yield of Corn, Washington, Mo., Weather Rev., Vol. 42, 1914, pp. 78–93.)

YIELD OF CORN IN BUSHELS PER ACRE

JULY PRECIPITATION IN INCHES	20.0 TO 24.9	25.0 TO 29.9	30.0 TO 34.9	35.0 TO 39.9	40.0 TO 44.9
80–89					1
70–79					
60–69			1	1	1
50–59			1	7	2
40–49	1	2	4	8	1
30–39		1	8	7	
20–29		5	5	1	1
10–19		1	1		

2. If it appears from this examination that a straight line is as good a fit as any other type of curve not too complicated

CORRELATION BETWEEN JULY PRECIPITATION AND YIELD OF CORN IN OHIO

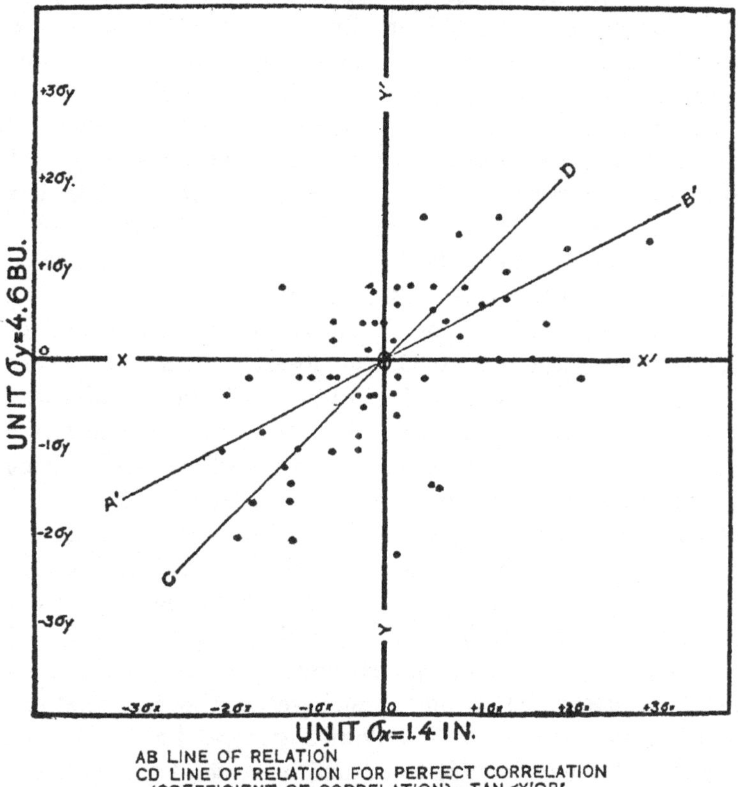

FIGURE 3.

to be useful as a measure of relationship, the data may be replotted on a new dot chart for which the unit of measurement on one axis is the standard deviation of one of the variables, and the unit on the other axis is the standard deviation of the other variable (see Figure 3).

3. The position of the straight line which most nearly satisfies the data on the second dot chart may be determined rigidly by the method of least squares. When the standard deviation of one variable is used as the unit of the ordinates and the standard deviation of the other variable as the unit of the abscissæ, the angles between this straight line of closest fit and the axis are significant. If these angles are equal, *i.e.* each 45°, the relationship between the variables is perfect (see *C–D* in Figure 3). If the line coincides with one axis or the other no relationship is shown, although the converse is not necessarily true.[1] Positions between these two show partial relationship (see *A'B'* in Figure 3).

4. The coefficient of correlation is merely a statement of the position of the straight line of closest fit on a chart where the units are the standard deviations of the variables as this position is determined by the least square adjustment.[2] The coefficient of correlation is expressed as the tangent of the angle made by the line of closest fit and the axis to which it is more nearly parallel (*e.g.* angle *X'OB'* in Figure 3 is $27\frac{1}{2}°$, tan $X'OB' = +0.526$). In actual practice the coefficient of correlation may be determined mathematically from the data as shown in Table I without plotting the material on a dot chart, like Figure 3. However, the coefficient should never be attempted without first investigating the relationship far enough to see if it follows a straight line. That is, steps 2 and 3 may be omitted in practice; step **1** should never be omitted.

5. If the examination of the correlation table or dot chart

[1] Yule, G. U.: Introduction to the Theory of Statistics, ed. 2, London, Griffin and Co., 1912, pp. 174–175.

[2] *Ibid.*, p. 172.

shows that the relation is not that of a simple straight line, the coefficient of correlation is not a measure of the relationship between the variables.

Limitations of the Coefficient of Correlation

It is clear even from a superficial study of the question that the coefficient of correlation obtained from material where a straight line relationship does not obtain may be too small,

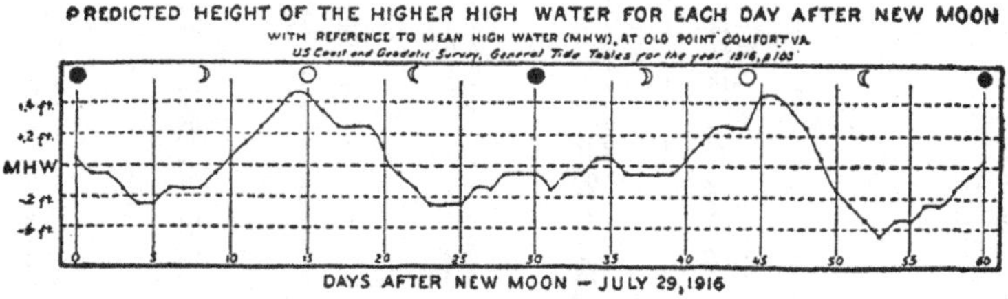

Figure 4.

but will never be too large.[1] A coefficient of correlation may be near zero when there is very close relationship, as is shown in such a condition as the relationship between the height of high water and the phase of the moon which is shown for Old Point Comfort, Va., by Table III and Figure 4. The figure indicates that the relation is harmonic; although there is a close and very definite relation between the phenomena, the coefficient of correlation is near zero ($-0.106 \pm .088$) because the different portions of the curve of regression are in such relations to each other that a straight line along an axis will most nearly satisfy all the points. Of course the angle is then zero and its tangent is zero.

[1] See Yule, G. U., "Introduction to the Theory of Statistics," ed. 2, London, Griffin & Co., 1912, p. 175, and Brown, W., "The Essentials of Mental Measurement," Cambridge, University Press, 1911, pp. 27–59.

Table III. Correlation of Time After New Moon and Predicted Height of the Higher High Water at Old Point Comfort, Va.

(U. S. Coast and Geodetic Survey, General Tide Tables for the Year 1916, p. 103.)

Days After New Moon, July 29, 1916	x	x^2	Height Above M. L. W.	y	y^2	xy
0	−30	900	2.7	+ .1	.01	− 3.
1	−29	841	2.6	0		
2	−28	784	2.6	0		
3	−27	729	2.5	− .1	.01	+ 2.7
4	−26	676	2.4	− .2	.04	+ 5.2
5	−25	625	2.4	− .2	.04	+ 5.0
6	−24	576	2.5	− .1	.01	+ 2.4
7	−23	529	2.5	− .1	.01	+ 2.3
8	−22	484	2.5	− .1	.01	+ 2.2
9	−21	441	2.6	0		
10	−20	400	2.7	+ .1	.01	− 2.0
11	−19	361	2.8	+ .2	.04	− 3.8
12	−18	324	2.9	+ .3	.09	− 5.4
13	−17	289	3.0	+ .4	.16	− 6.8
14	−16	256	3.1	+ .5	.25	− 8.0
15	−15	225	3.1	+ .5	.25	− 7.5
16	−14	196	3.0	+ .4	.16	− 5.6
17	−13	169	2.9	+ .3	.09	− 3.9
18	−12	144	2.9	+ .3	.09	− 3.6
19	−11	121	2.9	+ .3	.09	− 3.3
20	−10	100	2.7	+ .1	.01	− 1.0
21	− 9	81	2.6	0		
22	− 8	64	2.5	− .1	.01	+ .8
23	− 7	49	2.4	− .2	.04	+ 1.4
24	− 6	36	2.4	− .2	.04	+ 1.2
25	− 5	25	2.4	− .2	.04	+ 1.0
26	− 4	16	2.5	− .1	.01	+ .4
27	− 3	9	2.5	− .1	.01	+ .3
28	− 2	4	2.6	0		
29	− 1	1	2.6	0		
30	0	0	2.6	0		
31	+ 1	1	2.5	− .1	.01	− .1
32	+ 2	4	2.6	0		
33	+ 3	9	2.6	0		
34	+ 4	16	2.7	+ .1	.01	+ .4
35	+ 5	25	2.7	+ .1	.01	+ .5
36	+ 6	36	2.6	0		
37	+ 7	49	2.6	0		
38	+ 8	64	2.6	0		
39	+ 9	81	2.6	0		
40	+10	100	2.7	+ .1	.01	+ 1.0
41	+11	121	2.8	+ .2	.04	+ 2.2
42	+12	144	2.9	+ .3	.09	+ 3.6
43	+13	169	2.9	+ .3	.09	+ 3.9
44	+14	196	2.9	+ .3	.09	+ 4.2
45	+15	225	3.1	+ .5	.25	+ 7.5
46	+16	256	3.1	+ .5	.25	+ 8.0
47	+17	289	3.0	+ .4	.16	+ 6.8
48	+18	324	2.9	+ .3	.09	+ 5.4
49	+19	361	2.7	+ .1	.01	+ 1.9
50	+20	400	2.5	− .1	.01	− 2.0
51	+21	441	2.4	− .2	.04	− 4.2
52	+22	484	2.3	− .3	.09	− 6.6
53	+23	529	2.2	− .3	.16	− 9.2
54	+24	576	2.3	− .3	.09	− 7.2
55	+25	625	2.3	− .3	.09	− 7.5
56	+26	676	2.4	− .2	.04	5.2
57	+27	729	2.4	− .2	.04	− 5.4
58	+28	784	2.5	− .1	.01	− 2.8
59	+29	841	2.6	0		
60	+30	900	2.8	+ .2	.04	+ 6.0
		18910		−3.9 +6.9 +3.0	3.24	−25.1

$M'_x = 30$ $\qquad$ $M'_y = 2.6$

$M_x = 30$ $\qquad$ $M_y = 2.6 + \frac{3.0}{61} = 2.65$

$\Sigma x = 0$ $\qquad$ $\Sigma y = +.05$

$\Sigma x^2 = 18910$ $\qquad$ $\Sigma y^2 = 3.24$

$\sigma_x = \sqrt{\frac{18910}{61} - 0}$ $\qquad$ $\sigma_y = \sqrt{\frac{3.24}{61} - .003}$

$= 17.6$ $\qquad$ $= .22$

$$\Sigma xy = -25.1$$

$$r = \frac{\frac{-25.1}{61} - 0}{17.6 \times .22}$$

$$= -\frac{.411}{3.87}$$

$$= -.106 \pm E_r$$

$$E_r = .674\ \frac{1-(-.106)^2}{\sqrt{61}}$$

$$= .674\ \frac{1-.0112}{7.8}$$

$$= 0.674 \times 0.13$$

$$r = -0.106 \pm 0.088$$

When the relation is not linear the concomitant variation may be shown by the use of a "correlation ratio," which is simply a further development of the theory of correlation.[1]

It is, however, not the purpose of this paper to consider relationships shown by curves of a higher order than a straight line, as such correlations involve more complicated mathematical theory and also require many more observations to be significant.

[1] See Pearson, K., "Mathematical Contributions to the Theory o. Evolution," 14, on the general theory of skew correlation and non-linear regression. London, Drapers Company Research Memoirs. Biometric Series 2, 1905. Brown. W., "The Essentials of Mental Measurement," Cambridge, University Press, 1911, pp. 57–59.

Adequacy of the Coefficient of Correlation

The conclusion seems legitimate that the coefficient of correlation may be used strictly as a *measure* of relationship, when such relationship has been determined by other investigation to follow straight line relations. The use of the coefficient of correlation is to be recommended because it is independent of the personal equation of the investigator, and of the units employed, and because it shows rigidly the correct position of the line indicated by the dot chart.

In using the coefficient of correlation it is desirable to calculate the probable error (see Tables I and III for method).[1] The probable error is that divergence from the observed mean on either side within which half the observations lie. Its size is a measure of how closely the results from an infinite number of cases would correspond with those obtained from the observed cases. When the coefficient of correlation is not greater than its probable error there is no evidence that there is any correlation; but when the coefficient of correlation is clearly greater than its probable error correlation is indicated; and when it is much greater (six times as great is an accepted empirical amount) it may be safely assumed that there is concomitant variation.[2]

The coefficient of correlation is obtained by applying the least square adjustment to all the material and is, therefore, the straight line of closest fit. If the relationship is not that of a straight line, it is obvious that the straight line of closest fit is not a good measure of the relationship and that some other measure (*e.g.* the correlation ratio) must be used.

[1] For a general discussion of the significance of probable error see Yule, G. U., "Introduction to the Theory of Statistics," ed. 2, London, Griffin & Co., 1912, pp. 310–311.

[2] See Bowley, A. L., "Elements of Statistics," ed. 3, New York, Scribner, 1907, p. 320.

Therefore, the coefficient of correlation should never be used to show relationship until after the phenomena have been investigated, at least far enough to show whether a straight line satisfies the relationship as well as any other curve.

Literature

The development of the theory of correlation resulting in the adoption and use of the coefficient of correlation is, of course, largely mathematical. While the literature on the subject is considerable, the greater part of the contributions are concerned with the application of the coefficient to particular problems, and hence the development of the theory of correlation is incidental and widely scattered.

"The fundamental theorems of correlation were for the first time and almost exhaustively discussed by A. Bravais [1] . . . [more than] half a century ago. He deals completely with the correlation of two and three variables. Forty years later Mr. J. D. Hamilton Dickson [2] dealt with a special problem proposed to him by Mr. Galton, and reached on a somewhat narrow basis some of Bravais' results for correlation of two variables. Mr. Galton at the same time introduced an improved notation which may be summed up in the 'Galton Function' or coefficient of correlation. This indeed appears in Bravais' work, but a single symbol is not used for it. In 1892 Professor Edgeworth, also unconscious of Bravais' memoir, dealt in a paper on 'Correlated Averages' with correlation for three variables.[3] He obtained results identical

[1] *Analyse mathématique sur les probabilités des erreurs de situation d'un point.* Paris, Academie des Sciences, *Memoires presentés par divers savants.* Series 2, Vol. 9, 1846, pp. 255–332.

[2] Appendix to Galton, F., "Family Likeness in Stature," London, Royal Society, Proceedings, Vol. 40, 1886, pp. 63–73.

[3] London, *Philosophical Magazine,* Series 5, Vol. 34, 1892, pp. 190–204.

with Bravais, although expressed in terms of 'Galton's functions.'"[1]

The following publications contain complete statements of the later development:

PEARSON, KARL: Contributions to the mathematical theory of evolution; London, Royal Society, Philosophical Transactions, Series A, as follows:

1. On the dissection of frequency curves, Vol. 185, 1894, pp. 71–110.
2. Skew variations in homogeneous material, Vol. 186, 1895, pp. 343–414.
3. Regression, heredity, and panmixia, Vol. 187, 1896, pp. 253–318.
4. On the probable errors of frequency constants and on the influence of random selection on variation and correlation, Vol. 191, 1898, pp. 229–311.
5. On the reconstruction of the stature of prehistoric races, Vol. 192, 1898, pp. 169–244.
6. Genetic (reproductive) selection; inheritance of fertility in man and of fecundity in thoroughbred race horses, Vol. 192, 1899, pp. 257–330.
7. On the correlation of characters not quantitatively measurable, Vol. 195, 1900, pp. 1–47.
8. On the inheritance of characters not quantitatively measurable, Vol. 195, 1900, pp. 75–150.
9. On the principle of homotyposis and its relation to heredity, to the variability of the individual, and to that of the race, Vol. 197, 1901, pp. 285–379.
10. Supplement to a memoir on skew variation, Vol. 197, 1901, pp. 443–459.
11. On the influence of natural selection on the variability and correlation of organs, Vol. 200, 1902, pp. 1–66.
12. On a generalized theory of alternative inheritance with special reference to Mendel's Laws, Vol. 203, 1904, pp. 53–86.

[1] Pearson, Karl, *London Royal Society Philosophical Transactions*, Series A, Vol. 187, 1896, p. 261.

In London, Drapers' Company Research Memoirs, Biometric Series.

13. On the theory of contingency and its relation to association and normal correlation. Memoir 1.
14. On the general theory of skew correlation and non-linear regression. Memoir 2.
15. On the mathematical theory of random migration. Memoir 3, 1906.
16. On further methods of determining correlation. Memoir 4, 1907.
17. [Not published.]
18. On a novel method of regarding the association of two variates classed solely in alternate categories. Memoir 7, 1912.

PEARSON, KARL: On the partial correlation ratio. London, Royal Society, Proceedings, Series A, Vol. 91, 1915, pp. 492–498.

BROWN, W.: The essentials of mental measurement, Cambridge, University Press, 1911.

ELDERTON, W. P.: Frequency curves and correlation. London, Layton Brothers, 1906.

HOOKER, R. H.: Correlation of successive observations, Royal Statistical Society Journal, Vol. 68, pp. 676–703.

TOLLEY, H. R.: The theory of correlation as applied to farm survey data on fattening baby beef, U. S. Department of Agriculture Bul. 504, Washington, Govt. Ptg. Office, 1917.

WALKER, GILBERT T.: Correlation in seasonal variation of weather, Indian Meteorological Department Memoirs, Simla, 1909–1915.

1. Correlation in seasonal variation of climate, Vol. 20, part 6, 1909, pp. 117–124.
2. (A) On the probable error of a coefficient of correlation with a group of factors.
   (B) Some applications of statistical methods to seasonal forecasting, Vol. 21, part 2, 1910, pp. 22–45.
3. On the criterion for the reality of relationships or periodicities, Vol. 21, part 9, 1914, pp. 13–16.
4. Sunspots and rainfall, Vol. 21, part 10, 1915, pp. 17–60.
5. Sunspots and temperature, Vol. 21, part 11, 1915, pp. 61–90.
6. Sunspots and pressure, Vol. 21, part 12, 1915, pp. 91–118.

YULE, G. UDNY: Introduction to the theory of statistics, ed. 2, London, C. Griffin & Co., 1912, pp. 157–253.

More elementary discussions are contained in the following papers:

PERSONS, W. M.: The correlation of economic statistics. Boston, American Statistical Association, Quarterly Publications, Vol. 12 (1910), pp. 287–322.

HOOKER, R. H.: An elementary explanation of correlation: illustrated by rainfall and the depth of water in a well; London, Royal Meteorological Society Quarterly Journal, Vol. 34, 1908, pp. 277–291.

ELDERTON, W. P. and E. M.: Primer of statistics, London, A. and C. Black, 1910, pp. 55–72.

KING, W. I.: Elements of statistical method, New York, Macmillan, 1912, pp. 197–215.

DINES, W. H.: The practical application of statistical methods to meteorology. London, H. M. Meteorological Office, The computer's handbook (M. O. 223), section 5, part 2, 1915, pp. V29–V52.

The most complete bibliographies will be found in:

YULE, G. UDNY: Introduction to the theory of statistics, London, C. Griffin & Co., 1912, pp. 188, 208–209, 225–226, and 252.

DAVENPORT, C. B.: Statistical methods with special reference to biological variation, third, revised edition, New York, J. Wiley & Sons, 1914, pp. 62 and 85–104.

STATISTICAL STANDARDS IN THE INTERPRETATION OF FACTS[1]

Given a related group of statistical facts, having been collected, tabulated, and graphically expressed, to what standards must an interpretation of them conform? To fail to attach meaning and significance to them is simply to accentuate the all too prevailing practice of leaving untranslated

[1] Adapted from Secrist, Horace, "Statistical Standards in Business Research," *Quarterly Publications, American Statistical Association,* March, 1920, pp. 55–57.

into standards and principles the myriads of facts daily growing out of, or experienced in, human relations.

Certain fundamental standards of interpretation are the following:

First. The truth is the end sought: error is not to be disguised, falsehood tolerated, nor preconceptions favored.

Second. Comparisons can be made only between things, conditions, times, and places having common qualities.

Third. In interpretation, facts must always be referred to conditions which can produce them.

Fourth. Interpretation should extend to an explanation of the past and a forecast of the future.

Fifth. Distinction should be made between long- and short-time conditions and consequences; between transitory skirmishes and general tendencies.

Sixth. Distinction should be made between the result of a single cause and a combination of causes.

Seventh. Distinction should be made between drawing a particular deduction and giving it general application.

Eighth. Similarities and differences should be appraised in the light of particular application. Similarities which are seemingly complete and differences which are fundamental for one purpose may be ignored for others.

Ninth. The detail of interpretation should conform to the nature of the problem and the capacity of those interested. Not infrequently an exaggerated accuracy, which the nature of the basic data does not justify, nor the occasion for summarizing warrant, is worked out in detail by means of percentages, averages, and other summary expressions. Similarly, far-reaching conclusions are sometimes drawn from inadequate data by elaborate and overrefined methods. Statistical analysis then appears as an inverted and unstable pyramid.

Likewise, involved and complex interpretations are sometimes prepared for those who are statistically ignorant of refined processes or for those who are disinclined to follow or uninterested in pursuing an elaborate analysis. A statistical interpretation designed to influence executive action or to enlist administrative support is rarely, if ever, to be couched in the same language or to include the same detail, as one which is intended to serve the simple purpose of record. Consumers of statistics not only differ in their statistical interests but also in their statistical horizons.

REVIEW PROBLEMS

Given the following data showing the annual outlay and value of product realized by 51 farmers living near Dallas, Wisconsin, determine:

1. The coefficient of correlation and its probable error for outlay and value of product. Record all the steps in the process and all significant figures.

2. Given the data on page 420, showing the value of feed consumed and product produced by 26 registered cows of the same breed and under the same management, determine by the direct method for the two series, the coefficient of correlation and its probable error. Carefully record each step in the process and include in your presentation of method all significant figures. Use the nearest whole numbers — dollars — in all instances. (The arrangement of similar material in chapter 12 of the *Text* may be taken as a guide.)

What does the coefficient seem to show? Do you regard the data as adequate? Why? Is the coefficient significant according to the rule established by Bowley?

Annual Outlay and Total Value of Product on Fifty-one Farms near Dallas, Wisconsin [1]

Annual Outlay	Value of Product	Annual Outlay	Value of Product
$ 421	$1285	$ 563	$ 962
932	2649	620	1015
434	1143	1392	2259
293	727	715	1146
333	799	1165	1868
1683	3644	885	1410
1334	2844	764	1162
775	1646	1173	1778
1026	2165	440	686
1379	2895	1595	2358
1344	2533	1090	1602
961	2018	978	1435
1675	3473	1595	2165
1203	2472	1358	1878
1734	3619	1703	2339
983	2000	1018	1309
395	749	1505	1898
1618	3016	1492	1853
739	1361	1211	1496
881	1610	1103	1320
1266	2307	1095	1219
1124	1963	932	1009
1695	2909	1263	1348
1278	2192	742	759
894	1522	804	713
		1469	1131

[1] Data furnished by Professor H. C. Taylor, the University of Wisconsin.

VALUE OF FEED CONSUMED AND VALUE OF PRODUCT PER COW OF 26 REGISTERED COWS OF THE SAME BREED UNDER THE SAME MANAGEMENT.[1]

VALUE OF FEED CONSUMED	VALUE OF PRODUCT PER COW	VALUE OF FEED CONSUMED	VALUE OF PRODUCT PER COW
$99.83	$246.10	$98.93	$174.64
86.42	207.76	82.69	143.61
91.05	216.52	82.94	143.18
94.05	220.01	87.03	150.02
94.06	214.87	89.07	153.51
86.06	183.53	83.52	143.61
84.20	176.39	83.10	140.46
86.70	178.56	89.16	150.68
86.75	178.11	83.01	136.60
86.57	166.70	89.32	145.41
88.52	169.20	82.22	131.35
94.01	179.25	99.74	157.28
86.23	157.20	84.77	122.22

[1] Data furnished by Professor H. C. Taylor, the University of Wisconsin.

INDEX

Printed in the United States of America.

Made in the USA
Monee, IL
04 November 2026

67391553R10243